RADIATION SOURCE Use and Replacement

ABBREVIATED VERSION

Committee on Radiation Source Use and Replacement

Nuclear and Radiation Studies Board
Division on Earth and Life Studies

NATIONAL RESEARCH COUNCIL
OF THE NATIONAL ACADEMIES

THE NATIONAL ACADEMIES PRESS
Washington, D.C.
www.nap.edu

THE NATIONAL ACADEMIES PRESS 500 Fifth Street, N.W. Washington, DC 20001

NOTICE: The project that is the subject of this report was approved by the Governing Board of the National Research Council, whose members are drawn from the councils of the National Academy of Sciences, the National Academy of Engineering, and the Institute of Medicine. The members of the committee responsible for the report were chosen for their special competences and with regard for appropriate balance.

This study was supported by U.S. Nuclear Regulatory Commission under grant # NRC-04-06-069. Any opinions, findings, conclusions, or recommendations expressed in this publication are those of the author(s) and do not necessarily reflect the views of the organizations or agencies that provided support for the project.

International Standard Book Number-13: 978-0-309-11014-3
International Standard Book Number-10: 0-309-11014-9

Additional copies of this report are available from the National Academies Press, 500 Fifth Street, N.W., Lockbox 285, Washington, DC 20055; (800) 624-6242 or (202) 334-3313 (in the Washington metropolitan area); Internet, http://www.nap.edu.

Printed in the United States of America.

THE NATIONAL ACADEMIES

Advisers to the Nation on Science, Engineering, and Medicine

The **National Academy of Sciences** is a private, nonprofit, self-perpetuating society of distinguished scholars engaged in scientific and engineering research, dedicated to the furtherance of science and technology and to their use for the general welfare. Upon the authority of the charter granted to it by the Congress in 1863, the Academy has a mandate that requires it to advise the federal government on scientific and technical matters. Dr. Ralph J. Cicerone is president of the National Academy of Sciences.

The **National Academy of Engineering** was established in 1964, under the charter of the National Academy of Sciences, as a parallel organization of outstanding engineers. It is autonomous in its administration and in the selection of its members, sharing with the National Academy of Sciences the responsibility for advising the federal government. The National Academy of Engineering also sponsors engineering programs aimed at meeting national needs, encourages education and research, and recognizes the superior achievements of engineers. Dr. Charles M. Vest is president of the National Academy of Engineering.

The **Institute of Medicine** was established in 1970 by the National Academy of Sciences to secure the services of eminent members of appropriate professions in the examination of policy matters pertaining to the health of the public. The Institute acts under the responsibility given to the National Academy of Sciences by its congressional charter to be an adviser to the federal government and, upon its own initiative, to identify issues of medical care, research, and education. Dr. Harvey V. Fineberg is president of the Institute of Medicine.

The **National Research Council** was organized by the National Academy of Sciences in 1916 to associate the broad community of science and technology with the Academy's purposes of furthering knowledge and advising the federal government. Functioning in accordance with general policies determined by the Academy, the Council has become the principal operating agency of both the National Academy of Sciences and the National Academy of Engineering in providing services to the government, the public, and the scientific and engineering communities. The Council is administered jointly by both Academies and the Institute of Medicine. Dr. Ralph J. Cicerone and Dr. Charles M. Vest are chair and vice chair, respectively, of the National Research Council.

www.national-academies.org

COMMITTEE ON RADIATION SOURCE USE AND REPLACEMENT

THEODORE L. PHILLIPS (*Chair)*, University of California, San Francisco
EVERETT BLOOM, University of Tennessee, Knoxville
DAVID R. CLARKE, University of California, Santa Barbara
LEONARD W. CONNELL, Sandia National Laboratories, Albuquerque, New Mexico
ROBIN GARDNER, North Carolina State University, Raleigh
C. RICHARD LIU, University of Houston, Texas
RUTH MCBURNEY, Conference of Radiation Control Program Directors, Austin, Texas
ERVIN B. PODGORSAK, McGill University, Montréal, Québec, Canada
TOR RAUBENHEIMER, Stanford Linear Accelerator Center, Palo Alto, California
STEPHEN WAGNER, American Red Cross, Rockville, Maryland
DAVID L. WEIMER, University of Wisconsin at Madison

NUCLEAR AND RADIATION STUDIES BOARD LIAISON

SUSAN LANGHORST, Washington University in St. Louis, Missouri (September 2006 to December 2006)
ANDREW M. SESSLER, E.O. Lawrence Berkeley National Laboratory, Berkeley, California (January 2007 to August 2007)

Staff
MICAH D. LOWENTHAL, Study Director
FEDERICO SAN MARTINI, Program Officer
TRACEY BONNER, Program Assistant (October 2006 to March 2007)
MANDI M. BOYKIN, Senior Program Assistant (April 2007 to April 2008)
MARILI ULLOA, Senior Program Assistant (May 2006 to September 2007)

NUCLEAR AND RADIATION STUDIES BOARD

REVIEWERS

This report has been reviewed in draft form by persons chosen for their diverse perspectives and technical expertise in accordance with procedures approved by the National Research Council's Report Review Committee. The purposes of this review are to provide candid and critical comments that will assist the institution in making the published report as sound as possible and to ensure that the report meets institutional standards of objectivity, evidence, and responsiveness to the study charge. The review comments and draft manuscript remain confidential to protect the integrity of the deliberative process. We wish to thank the following for their participation in the review of this report:

Ahmed Badruzzaman, Chevron Corp. and University of California, Berkeley
Anthony Berejka, Ionicorp+
Joel Bedford, Colorado State University
Barbara Bierer, Brigham and Women's Hospital
Paul Fischbeck, Carnegie Mellon University
Gregory Van Tuyle, Los Alamos National Laboratory
Karl Hemmerich, Steris Isomedix, Inc.
William Hendee, Wisconsin College of Medicine
Joel Lubenau, private consultant
Richard Meserve, Carnegie Institution of Washington and Nuclear and Radiation Studies Board of the National Research Council
John Poston, Texas A&M University
Bruce Thompson, Iowa State University
Raymond Wymer, Oak Ridge National Laboratory (retired)

Although the reviewers listed above have provided many constructive comments and suggestions, they were not asked to endorse, nor did they see the final draft of, the report before its release. The review of this report was overseen by Louis Lanzerotti, New Jersey Institute of Technology, and Thomas Budinger, University of California at Berkeley. Appointed by the National Research Council, they were responsible for making certain that an independent examination of this report was carried out in accordance with institutional procedures and that all review comments were carefully considered. Responsibility for the final content of this report rests entirely with the authoring committee and the National Research Council.

PREFACE

On September 11, 2001, terrorists turned passenger airliners in New York, Washington, and Pennsylvania into weapons, killing not only those onboard, but also thousands of people working in office buildings and many of those who tried to rescue them. The perverse success of those attacks has forced the nation to contemplate the possibility that other technologies that were designed, and are used, solely for the benefit of society—to treat and cure illness, to prevent infection, illness, and disease, to improve safety in industrial equipment, and to help obtain resources that we rely on every day—could be used maliciously against us. The U.S. Congress requested this study as part of a set of actions to improve the nation's security against attacks that might use radiation sources. The committee had this in mind throughout the study.

The committee's charge from the Congress was to evaluate technologies and make recommendations on options for implementing those technologies. Risk considerations were at the center of this task: What are the high-risk sources? What makes them hazardous? Would implementing replacements reduce risks? So, as the committee set out to hear from experts on the radiation source applications, the committee also examined the hazards associated with the radiation sources, including both accidents and malevolent acts.

Policy makers will seek to balance among alternative technologies, physical security measures, tracking and accounting, and intelligence and law enforcement operations to prevent or mitigate the consequences of a radiological attack. In this report, the committee offers its recommendations on replacements for high-risk radiation sources, including priorities among the sources, and options for implementing those replacements.

In carrying out the study, the committee was cognizant of the broad array of people and institutions that use and benefit from the use of radionuclide radiation sources, and that would be affected by the committee's recommendations. The committee met with practitioners and researchers in the relevant fields, talked with radionuclide source manufacturers and providers of alternatives or replacements for the sources, and visited facilities that use the radiation sources. The committee also talked, to some extent, to the customers for the services provided. The recommendations in this report, and indeed any moves for replacement of radionuclide radiation sources, will affect those people and institutions to varying degrees and in different ways. It is the committee's hope that any actions for implementation of replacements will consider the input from the people and institutions affected, because those people and institutions are providing important services to society, have a stake in the outcomes, and are likely to benefit or suffer the most from the government's actions.

Readers who examine this report thoroughly will notice that the task statement explicitly requests an evaluation of worker hazards from technologies meant to replace the current high risk radiation sources, and the report discusses these hazards only briefly (see, e.g., Chapter 4 and Chapter 6). This is not an oversight. The committee devoted little of the report to this topic because the most common replacement technologies, electron accelerators, and the high-risk radiation sources pose similar radiation hazards while the accelerator is operating, and the accelerators operating at energies below 10 MeV pose fairly insignificant radiation hazards when they are not operating. Further, these matters are already covered well in other reports (see especially NCRP, 2005). Readers might desire to see more on the costs and time lines for readiness of replacement technologies. These are discussed for many, but not all, replacements. In this case, more extensive discussion and detail were not included because the data available to the committee do not support saying more.

Throughout the study the committee benefited from the input of people who manufacture, use, and regulate the radiation source devices and their potential replacements. The committee thanks Bob Adolph, Mike Ault, Lester Boeh, Les Braby, Kevin Brooks, Ian Brown, Greg Budner, Tom Chadwick, Marshall Cleland, David Coppell, Mike Creech, Kirsten Cutler, Jim Dempsey, Donny Dicharry, Brian Dodd, Patricia Eifel, James Elrod, Hugh Evans, Tara Federici, Michal Freedhoff, Peter Fundarek, Colette Germain, Allen Gilchrist and Baker Hughes Incorporated, Michael Gillin, Paul Gray, Dave Hall, Sally Hamlin, Barbara Hamrick, Patricia K. Holahan, Merri Horn, Bob Irwin, Ramzi Jammal and the Canadian Nuclear Safety Commission, Carol Jantzen, Slobodan V. Jovanovic, Randol Kirk, Ken Koziol, Steve Laflin, Ann Lawyer, Benjamin Lichtiger, Glenn Light, Mick Lord, Grant Malkoske and MDS Nordion, John Masefield, Joseph E. Maxim, Ray Meyn, Radhe Mohan, Aaron Morrison, Paul Moses, Boris Myasoedov, Wayne Norwood, Pearce O'Kelley, Mike Pearson, Vladimir Pet'kov, Brendan Plapp, Jay Poston, Karl Prado and the M. D. Anderson Medical Center, André Régimbald, Robert Rushton, Ward Schultz, J. L. and Mary Shepherd, Mark Shilton, Almon Shiu, Mark Smith and Sterigenics, Inc., Mickey Speakmon, Makuteswara Srinivasan, Chris Stoller, Orhan Suleiman, Peggy Tinkey, David Tiktinsky, Chuck Vecoli, Mark Vist, Bill Ward, Tom Wasiak, Ruth Watkins, Richard Wiens, Shiao Woo, and Otto Zeck and Memorial Hermann Hospital.

These people were generous with their time, information, and advice. The committee would specifically like to acknowledge the support provided by the U.S. Nuclear Regulatory Commission (U.S. NRC) staff, especially the committee liaison, Tony Huffert, and his office director, Brian Sheron, who made efforts throughout the study to ensure that the U.S. NRC provided what it could to assist the committee in fulfilling its task. Finally, the committee thanks its staff: Mandi Boykin, Tracey Bonner, and Marili Ulloa were responsible for the care and feeding of the committee; Kevin Crowley and Federico San Martini made important contributions at key points in the study; and Micah Lowenthal provided the guidance, coordination, and various kinds of support the committee needed to get the job done well. They were all important to the successful completion of the study, and we are grateful to them for their help and support.

Prior to public release, and as required under the terms of the grant for this study, the report was sent to the U.S. NRC in August 2007 for security classification review. The agency determined that the report contained information that is exempt from public release under 5 U.S.C. § 552(b). In late January 2008, the National Research Council and the U.S. NRC reached agreement that this abbreviated version could be released to the public without restriction. The findings and recommendations remain substantively unchanged from the full version, which has been provided to the government. During the security classification review the U.S. NRC also provided additional non-security-related comments on the report. The National Research Council made some factual corrections and revised wording to improve clarity in the report in response to those comments (but made no other substantive changes) and consulted with its Report Review Committee about the nature of these changes.

While this abbreviated report was being readied for release, the committee was informed that the U.S. NRC has begun exploring options to address some of the concerns raised in the report. Although the committee has not received detailed information about the U.S. NRC's actions, the committee commends the U.S. NRC for these explorations and encourages the government to take steps that will facilitate replacement of high-risk radiation sources and improve radiation source safety and security.

Theodore L. Phillips, Chair
Committee on Radiation Source Use and Replacement

CONTENTS

EXECUTIVE SUMMARY

The U.S. Congress asked the National Research Council to review the civilian uses of radionuclide radiation sources and potential replacements for sources that pose a high risk to public health or safety in the event of an accident or attack. Considering technical and economic feasibility and risks to workers, the committee was asked to make findings and recommendations on options for implementing the identified replacements. In carrying out its charge, the committee met with practitioners and researchers in the relevant fields and, in this report, has focused foremost on hazards and risks, feasibility of replacements, and options for implementing the replacements.

Approximately 5,000 devices containing nearly 55,000 high-activity radiation sources are licensed for use today in the United States. The devices are used for cancer therapy, sterilization of medical devices, irradiation of blood for transplant patients and of laboratory animals for research, nondestructive testing of structures and industrial equipment, and exploration of geologic formations to find oil and gas deposits. These radiation sources and devices are licensed and regulated by the U.S. Nuclear Regulatory Commission (U.S. NRC) or by state agencies with authority to regulate materials covered by agreements with the U.S. NRC (Agreement States). Because the array of applications of these radiation sources is so broad and the applications are essential to securing health, safety, and prosperity, the devices are licensed for use and found in every state in the nation. Some types of radiation sources should be replaced with caution, ensuring that the essential functions that they perform are preserved.

For prioritizing its efforts to reduce security risks, the U.S. NRC should consider radiation sources' potential to cause contamination of large areas resulting in area denial.

Out of the thousands of manufactured and natural radionuclides, americium-241, cesium-137, cobalt-60, and iridium-192 account for nearly all (over 99 percent) of the sealed sources that pose the highest security risks in the United States. Of the radionuclides mentioned above, cesium-137 in the form of cesium chloride is a greater concern than other radiation sources based on its dispersibility and its presence in population centers across the country. In view of the overall liabilities associated with radioactive cesium chloride and the alternatives that are available now or possible in the future to replace these radiation sources, the committee finds that high-activity cesium chloride sources should be replaced. The committee suggests options for implementing the replacement, including discontinuation of licensing of new cesium chloride irradiator sources and devices and incentives to decommission existing sources and devices. In addition, the committee finds that nonradionuclide replacements exist for nearly all applications of the radiation sources examined, but they may not all now be economically viable or practical. Neither licensees nor manufacturers now bear the full life-cycle cost, including disposal costs, of some of these radiation sources. The committee recommends that the U.S. government provide incentives (market, regulatory, and certification) to facilitate the introduction of alternatives for the high-risk radiation sources and reduce the sources' attractiveness and availability. These and related findings and recommendations are discussed in detail in the body of the report.

The study task did not include detailed cost-benefit analyses and did not permit examination of lower activity radiation sources (Category 3 or lower), even in aggregation.

SUMMARY

Several thousand devices containing nearly 55,000 high-activity[1] radiation sources are licensed for use today in the United States. The devices are used for cancer therapy, sterilization of medical devices, irradiation of blood for transplant patients and of laboratory animals for research, nondestructive testing of structures and industrial equipment, and exploration of geologic formations to find oil and gas deposits. These radiation sources and devices are licensed and regulated by the U.S. Nuclear Regulatory Commission (U.S. NRC) or by state agencies with authority to regulate materials covered by agreements with the U.S. NRC, called Agreement States. Because the array of applications of these radiation sources is so broad and the applications are essential to securing health, safety, and prosperity, the devices are licensed for use and found in every state in the nation.

After the terrorist attacks on the United States on September 11, 2001, concerns about the safety and security of these radiation sources and devices grew, particularly amid fears that terrorists might use radiation sources to make a radiological dispersal device or "dirty bomb." As part of the Energy Policy Act of 2005, the U.S. Congress directed the U.S. NRC to take several actions, including requesting a study by the National Research Council to identify the legitimate uses of high-risk radiation sources and the feasibility of replacing them with lower risk alternatives. The committee appointed by the National Research Council to carry out the study was tasked to provide a review of radiation source use, potential replacements for sources that pose a high risk to public health or safety, and findings and recommendations on options for implementing the identified replacements. To do that, the committee met with practitioners and researchers in the relevant fields, examined scientific research and trade literature, and visited facilities that use the radiation sources.

In carrying out its charge, the committee has focused foremost on hazards and risks,[2] feasibility of replacements, and options for implementing the replacements. This study is not the first effort to examine the uses for radionuclide radiation sources and prioritize among them based on certain kinds of risk. A number of studies (see, e.g., Ferguson et al., 2003; Van Tuyle et al., 2003) describe the system of supply of radionuclide radiation sources and their applications. The Department of Energy (DOE) and the U.S. NRC issued a joint report identifying risk-significant radiation sources and quantities of radioactive material (DOE/U.S. NRC, 2003). The IAEA, in a similar but broader effort, revised its Code of Conduct on the Safety and Security of Radioactive Sources (2003), which provides guidelines for countries in the development and harmonization of policies, laws, and regulations on the safety and security of radioactive sources. The IAEA Code of Conduct includes a categorization system for radionuclide radiation sources that provides a risk-based ranking of radioactive sources based on their potential for harm to human health under specific scenarios and for grouping of source use practices into discrete categories. The radiation sources in Category 1 are those that, if not managed safely or securely, could lead to the death or permanent injury of individuals in a short period of time. Similarly, Category 2 sources are those that could lead to the death or

---

[1] Activity is the number of radioactive decays per second. Specific activity is the activity per gram of material. The high-activity sources cited here are Category 1 and 2 sources, as defined in the International Atomic Energy Agency's (IAEA's) Code of Conduct on the Safety and Security of Radioactive Sources, and described in this summary.

[2] For clarity, and to be consistent with the standard scientific definitions, the committee uses the term *hazard* to denote the potential to cause harm and the term *risk* to describe a hazard linked to a context of exposure or possibility of an event leading to exposure. Gasoline is hazardous; gasoline stored where an open flame or spark might ignite it poses a high risk.

permanent injury of individuals who may be in close proximity to the radioactive source for a longer period of time than for Category 1 sources. Based on direction and authority in the Energy Policy Act of 2005 (P.L. 109-58), the U.S. NRC limited the radiation sources within the study scope to Category 1 and 2 sources.

Data from the U.S. NRC show that out of the thousands of manufactured and natural radionuclides, americium-241, cesium-137, cobalt-60, and iridium-192 account for nearly all (over 99 percent) of the Category 1 and 2 sources. The features of these and some other key radionuclide radiation sources are summarized in Table S-1.

TABLE S-1 Summary of Radionuclides in Category 1 and 2 Radiation Sources in the United States[a]

Radionuclide	Half-life	Radioactive Emissions and Energies	Typical Specific Activity (TBq/g) [Ci/g]	Total Activity in U.S. Inventory (TBq) [Ci]	Major Applications	Typical Activity (TBq) [Ci]	Physical or Chemical Form
Americium-241	432.2 yr	α–5.64 MeV, γ-60 keV, principal	0.13 [3.5]	240 [6,482]	Well logging	0.5–0.8 [13–22]	Pressed powder (americium oxide)
Californium-252	2.645 yr	α–6.22 MeV, fission fragments, neutrons, and gamma rays	20 [540]	0.26 [7]	Well logging	0.0004 [0.011]	Metal oxide
Cesium-137 (Ba-137m)	30.17 yr	β-518 keV max with γ-662 keV (94.4% of decays) or β-1.18 MeV max	0.75 [20]	104,100 [2.8 million]	Self-contained irradiators Teletherapy Calibrators	75 [2,000] 50 [1,400] 15 [400]	Pressed powder (cesium chloride)
Cobalt-60	5.27 yr	γ-1.173 and 1.333 MeV	3.7 [100]	7.32 million [198 million]	Panoramic irradiators Self-contained irradiators	150,000 [4 million] 900 [24,000]	Metal slugs
			11 [300]		Teletherapy Industrial radiography	500 [14,000] 4 [100]	Metal pellets
Iridium-192	74 d	β-1.46 MeV max with 2.3 γ-380 keV avg, 1.378 MeV max (0.04% of decays)	18.5 [500]	5,436 [146,922]	Industrial radiography	4 [100]	Metal
Plutonium-238	87.7 yr	α–5.59 MeV, and γ-43 keV (30% of decays)	2.6 [70]	34.7 [937]	RTG Pacemakers (obsolete) Fixed gauges	10 [270] 0.1 [3] 0.75 [20]	Metal oxide
Selenium-75	119.8 d	γ-280 keV average, 800 keV max	20–45 [530–1200]	9.7 [261]	Industrial radiography	3 [75]	Elemental or metal compound
Strontium-90 (Yttrium-90)	28.9 yr	β-546 keV	5.2 [140]	64,000 [1.73 million]	RTG	750 [20,000]	Metal oxide

[a] Nuclear decay data for this table and throughout the report are from Firestone and Shirley (1996).

Consideration of technological alternatives to radionuclide radiation sources has been recommended by the Health Physics Society, the IAEA, and others. The replacement options may include replacing the radionuclide-based technology with a technology not involving radiation or with x-rays, an electron beam, or neutrons from a radiation generator (a particle accelerator device). Finally, the radionuclide or the chemical and physical form of the radionuclide may be changed to a less hazardous one.

In the body of the report the committee discusses origins, forms, and applications of radionuclide radiation sources (Chapter 2), risks associated with radionuclide radiation sources (Chapter 3), accelerator and detector technologies (Chapter 4), each of the major applications of radionuclide radiation sources (Chapters 5 through 9), and options for implementation of application-specific replacement technologies, including the various kinds of incentives that might be applied (Chapter 10).

The major findings and recommendations are described below and are discussed in detail in the body of the report.

FINDINGS AND RECOMMENDATIONS

Finding 1: The radiation sources examined in this study are used in applications that are important to the nation's health, safety, and economic strength.

High-activity radiation sources are used in the United States and other modern societies in a variety of ways: They are used in devices that improve the success of medical procedures—ensuring that medical devices and implants are sterile, preventing fatal complications from bone marrow transplants, and providing noninvasive techniques for treating brain lesions; they are used in devices for inspecting the integrity of buildings, bridges, and industrial equipment; and they are used to seek out oil and gas resources deep in the ground. These applications are immensely valuable to the United States. The question is not whether these activities should continue, but whether lower risk replacements for the radiation sources are feasible and practical, and what steps should be taken to implement replacements for the sources that pose a high risk to public health and safety

Recommendation 1: Replacement of some radionuclide radiation sources with alternatives should be implemented with caution, ensuring that the essential functions that the radionuclide radiation sources perform are preserved.

As the nation seeks to improve safety and security, the value and benefits of current practices should be recognized and, where possible, the services the devices provide should not be compromised. Some replacements do entail trade-offs with respect to safety, security, costs, convenience, and performance, as discussed in Chapters 3 through 9. These trade-offs should be considered carefully. A reduction in the performance of a device may be acceptable if it provides sufficient benefits in safety, for example. Replacement should preserve acceptable performance of these applications to preserve the benefits that these applications provide, on many of which the United States has come to rely.

Finding 2a: The U.S. NRC ranks the hazards of radiation sources primarily based on the potential for deterministic health effects (especially death and severe bodily harm) from direct exposure to the radiation emitted by the bare (unshielded) sources. The U.S. NRC's analyses that support the commission's security requirements for nuclear materials licensees are based only on these potential consequences.

The U.S. NRC has ranked radiation sources in terms of hazard using the IAEA system of five source categories, determining that the Category 1 and 2 sources are "high-risk sources." The IAEA analyses supporting its source categorization system consider only deterministic health effects (such as early fatalities) from direct exposure to ionizing radiation from the unshielded source under different exposure scenarios. The initial DOE/U.S. NRC analysis used the same consequences and added a contamination threshold criterion that does not account well for the differing potential for area denial or economic consequences of dispersal attacks with different radiation sources. The U.S. NRC also carried out security analyses of each type of facility licensed to use Category 1 and 2 sources, but these analyses were confined to examining the potential for deterministic health effects caused by attacks involving the Category 1 and 2 sources. The U.S. NRC staff told the committee that this was seen as a first step, and that the commission was considering whether to include other factors.

Finding 2b: Factors other than the potential to cause deterministic health effects are important when evaluating hazards from radiation sources, especially the potential to cause contamination of large areas resulting in economic and social disruption (area denial).

A radiological incident (an accident or especially an attack) could have its most long-lasting and far-reaching effects as a result of contamination of land, buildings, and infrastructure in densely populated regions, partially or completely disabling those assets for human use for long periods of time. This is illustrated by the radiotherapy source incident in 1987 in Goiania, Brazil, and the Chornobyl nuclear reactor accident in the Ukraine. Although an event like the Chornobyl reactor fire is not possible with radiation sources and the scale of the contamination from an incident with radiation sources would inherently be smaller, that 1986 accident showed that radioactive contamination can create sizeable areas that are deemed uninhabitable for extended periods of time. The economic and social disruptions caused by such incidents can be difficult to quantify, but they are critical to understanding the scope of the impact beyond the fatalities and severe bodily injuries caused by these events.

Recommendation 2: For prioritizing efforts to reduce risks from malicious use of radiation sources, the U.S. NRC should consider radiation sources' potential to cause contamination of large areas resulting in economic and social disruption (area denial) to determine what, if any, additional security measures are needed.

Having taken an essential first step in considering deterministic health effects from possible radiation exposure from an incident involving radiation sources, the U.S. NRC should now include economic and social disruption in its risk analyses of radiation sources. These impacts can vary significantly depending on the scenarios considered, but that variability does not make them less important. Further, even with such variability, certain factors emerge as important in other analyses of these issues (e.g., Van Tuyle et al., 2003). In carrying out its analyses, the U.S. NRC should not confine itself to the numeric source-activity cutoffs defining the lower limits for Category 1 and 2 sources because the source categorization system itself is based on deterministic health effects. For example, many self-contained irradiators are Category 2 devices, but are near the Category 1 threshold, and most americium-beryllium well logging sources have activities near but below the Category 2 limit. Review may show that each set of these devices should be regulated similarly.

After 2001, the U.S. NRC imposed enhanced security requirements on its materials licensees: Compensatory Measures for panoramic irradiators, Additional Security Measures for its manufacturers and distributors, and Increased Controls for licensees with Category 1 and 2 devices and sources. Compensatory Measures include fairly robust access controls and alarms with response by armed security personnel, along with other measures. Increased Controls include access controls and alarms with response by security personnel, and other measures. After review of the risks associated with some sources and devices considering more fully the potential for contamination from an attack, the U.S. NRC might conclude that more stringent measures are needed for some Category 1 and 2 sources and devices. The committee did not examine these security matters in detail and so cannot prejudge the outcome of such analyses. The committee does note, however, that such measures could improve the security of the devices and create a disincentive for owning them.

Finding 3a: Because of its dispersibility, solubility, penetrating radiation, source activity, and presence across the United States in facilities such as hospitals, blood banks, and universities, many of which are located in large population centers, radioactive cesium chloride is a greater concern than other Category 1 and 2 sources for some attack scenarios. This concern is exacerbated by the lack of an avenue for permanent disposal of high-activity cesium radiation sources, which can result in disused cesium sources sitting in licensees' storage facilities. As such, these sources pose unique risks.

Radioactive cesium chloride sources are in the form of a steel-encapsulated, compressed powder. The salt is highly dispersible and water soluble. There are approximately 1,300 high-activity cesium chloride devices (each with an activity of tens to hundreds of terabecquerels [hundreds to thousands of curies]) across the United States, nearly all of which are self-contained irradiators. The number of these devices and sources appears to be increasing.

Because it emits energetic gamma rays and its half-life is long enough that an irradiator does not need to be reloaded over the device's expected lifetime, cesium-137 has been the key component of self-contained irradiators for blood irradiation and research for many years. Cesium chloride is the least expensive and highest-specific-activity form of cesium-137 available today. Because of the nature of the applications that employ these irradiators, they are most commonly located in hospitals, blood banks, and universities, many of which are located in cities, large and small. The presence of these sizable sources in areas that are potentially attractive targets is a major factor making radioactive cesium chloride such a concern to the committee.

Finding 3b: In view of the overall liabilities of radioactive cesium chloride, the committee judges that these sources should be replaced in the United States and, to the extent possible, elsewhere.

Finding 3c: In most (and perhaps all) applications, radioactive cesium chloride can be replaced by (1) less hazardous forms of radioactive cesium, (2) radioactive cobalt, or (3) nonradionuclide alternatives. However, not all of these alternatives are commercially

available now, and all are currently more expensive than radioactive cesium chloride for the users.

Some alternatives to radioactive cesium chloride include radioactive cesium glass and a mineral form (pollucite) loaded with radioactive cesium (described in Chapter 2). These alternative material forms use the same cesium-137 as radioactive cesium chloride; thus the gamma rays and the half-life are identical, but the specific activity of these sources is smaller and the pollucite is more difficult to fabricate, especially for high-activity sources. The committee judges that none of the current applications of high-activity cesium sources about which the committee was informed necessitates the higher specific activity afforded by cesium chloride. Accommodating the larger volume needed to achieve the same source activity would require redesign of some (but not all) devices. High-activity cesium sources are not, however, available in these alternative material forms today, and making them available may require the cesium source producer (the Production Association Mayak, in Russia) to modify its production process.

Cobalt-60 may be substituted for radioactive cesium chloride for many applications (see the discussion in Chapter 5), although as much as twice the shielding thickness may be required for a source that achieves the same dose rate, and the half-life of cobalt-60 is shorter (5.3 years for cobalt-60 versus 30 years for cesium-137), thus lowering significantly the useful lifetime of the source. Shielding challenges can be addressed in part by switching from lead shields to more effective tungsten or depleted uranium shielding, but tungsten shielding is more expensive than lead and manufacturing depleted uranium shielding is a very specialized, expensive operation that requires U.S. NRC licensing for its whole life cycle. The shorter useful lifetime of cobalt-60 radiation sources requires that they be replaced periodically, which entails transportation of a fresh source and the used source, with the attendant risks associated with source transportation.

X-ray generators are already commercially available as substitutes for applications that do not require the gamma rays with definite energies emitted by cesium-137 and cobalt-60. X-ray tubes can be expensive and require more maintenance for periodic calibration and replacement than radioactive sources require.

Finding 3d: Government action is required to implement replacement of radioactive cesium chloride sources because the alternatives cost more and the liabilities or social costs of the sources currently are not borne by the end users.

There is no indication that replacement of devices containing Category 1 and 2 radioactive cesium chloride sources with lower hazard alternatives will improve or worsen the performance of the devices in their standard and proper uses. The act of replacement incurs monetary costs, and the replacements themselves currently cost more in most cases than the radioactive cesium chloride devices. All of these costs would be borne by the end users (paying more for the alternatives) and the current device manufacturers (depending on the price elasticity of demand and potential loss of sales). The benefits of replacement are in reducing the liabilities and social costs (including the costs associated with the risk of terrorist attacks and, in some cases, the full costs of disposal, discussed in Chapters 2 and 10). Those social costs, including the risks, are shared by the public. Except in cases where the replacements prove to be cheaper, end users have little incentive to shift away from radioactive cesium chloride; and unless there is a demand for the alternatives, manufacturers are unlikely to invest in making the alternatives available. Government action can, however, provide the requirements or incentives to implement replacement.

Recommendation 3: In view of the overall liabilities of radioactive cesium chloride, the U.S. government should implement options for eliminating Category 1 and 2 cesium chloride sources from use in the United States and, to the extent possible, elsewhere. The committee suggests these options as the steps for implementation:

- i. **Discontinue licensing of new cesium chloride irradiator sources.**
- ii. **Put in place incentives for decommissioning existing sources.**
- iii. **Prohibit the export of cesium chloride sources to other countries, except for purposes of disposal in an appropriately licensed facility.**

In Chapter 10, the committee offers several suggestions as its lead candidates for how to implement the replacement, and they are summarized here. First, to stop the addition of new Category 1 and 2 cesium chloride sources to the nation's inventory, the U.S. NRC should discontinue all new licensing and importation of these sources and devices. This includes import of new sources from other countries and recycling of sources from decommissioned devices. Second, many licensees may need incentives to decommission their existing sources or devices because the devices still have use value. Indeed, there are now also disincentives to decommissioning beyond the loss of use, including the costs of decommissioning. Third, if the sources recovered from decommissioned devices (or the devices themselves) are simply sold outside the United States, then the sources are still potentially available for use in an attack on another country or even the United States. Therefore, disposition options are needed in the United States. These are discussed in more detail in Chapter 10.

The overall policy could make exceptions based on unique needs that cannot be met with alternative technologies, but the threshold for creating exceptions should be set high, similar to what the U.S. NRC has done for panoramic irradiators.

Finding 4a: Nonradionuclide replacements exist for nearly all applications of Category 1 and 2 radionuclide sources (not just radioactive cesium chloride). At this time, these replacements may not all be practical or economically attractive, but most of them are improving.

Chapter 4 shows a variety of accelerator systems that can be designed to operate as radiation-generator replacements for radionuclide radiation sources. In Chapter 5, the committee explains that *self-shielded irradiators* can be operated with x-ray generators instead of radionuclides. Some x-ray-based irradiators are already commercially available and more companies that design and manufacture x-ray generators told the committee that they are considering entering the market. As described in Chapter 6, large companies in the business of *sterilization of medical supplies and devices* operate several kinds of facilities (ethylene oxide, gamma irradiation, and electron beam irradiation) to use the technology that is best suited to the sterilization contract. An x-ray irradiation facility can be a direct replacement for a cobalt-60 panoramic gamma irradiator, and may offer both electron-beam and x-ray irradiation in one facility. Some supporters of x-ray irradiation have concluded that larger x-ray facilities (several hundred kilowatts) would have economic advantages. The first of these larger scale facilities for x-ray irradiation is to be built soon in Belgium. It is unclear whether such facilities will be cost-neutral, more expensive, or less expensive per pallet irradiated than similarly sized gamma irradiators. As noted in Chapter 7, linear accelerators for *radiotherapy* have almost entirely replaced cobalt-60 teletherapy devices in the United States, except for the Gamma Knife®, the use of which is still growing. The Gamma Knife® is less versatile than a linear accelerator for radiotherapy, but offers features that some customers perceive to be advantages, which their competitors are trying to match with accelerators. The development of new technologies, especially in the areas of ultrasonics and x-ray sources, has provided several alternatives to *gamma radiography* in the field of nondestructive inspection. In some areas, it is likely that the use of some of the alternatives is currently limited by the availability of trained personnel and wider acceptance of the results as durable records of proper inspection, as noted in Chapter 8. Chapter 9 similarly explains that the *neutron well logging* tools that use americium-beryllium sources are beginning to see competition from accelerator fusion sources. Table S-2 summarizes the radiation source applications and replacements.

Finding 4b: Neither licensees nor manufacturers now bear the full cost of liabilities related to misuse of Category 1 and 2 radiation sources, nor do they bear the costs of disposal of cesium and americium sources.

Category 1 and 2 radiation source licensees are not required to be insured for the possible consequences of a malicious use of their radiation sources. This is no different than in other sectors of our society, but it means that the costs of some liabilities are not borne by licensees. In addition, licensees of Category 1 and 2 cesium-137 and americium-241 sources in the United States do not now bear the costs of disposal of their sources because the disposal facilities for these high-activity sources can only accept sources that come from DOE or its predecessor, the Atomic Energy Commission. DOE has a program called the Offsite Source Recovery Project, which packages, transports, and stores high-risk radiation sources and devices without fee. Some licensees pay for the cost of packaging and transportation to effect the removal on their own schedule, but the cost is lower than the cost of disposal will be in an as-yet-unknown disposal facility for "Greater than Class C" low-level waste.

Recommendation 4: In addition to actions related to radioactive cesium chloride, the U.S. government should adopt policies that provide incentives (market, regulatory, or certification) to facilitate the introduction of replacements and reduce the attractiveness and availability of high-risk radionuclide sources.

The committee describes several options for implementation of alternatives in Chapter 10 of the report. Among these options are to make licensees bear the full life-cycle cost of radiation sources, particularly for disposal of cesium-137 and americium-241 sources; to revise the requirements for decommissioning funds for Category 1 and 2 devices to increase the up-front costs for higher hazard sources; enhance the DOE's Offsite Source Recovery Project to include a buyback of devices that still have use value, provided that the devices are replaced with lower hazard devices. The government could impose charges on all sources, or just on new sources, based on hazards or risks.

Finding 5: Accelerator neutron sources and californium-252 sources show promise as potential replacements for americium-beryllium sources in neutron well logging tools. However, there are technical obstacles for these replacement sources and they are at a disadvantage based on the extensive experience and data accumulated with americium-beryllium sources.

Recommendation 5: The Society of Petrophysicists and Well Log Analysts should task an industry working group, called a Special Interest Group (SIG) such as the Nuclear Logging SIG, to address the technical obstacles to implementing replacements for the

americium-beryllium sources used in well logging and the challenges of data interpretation. The group should decide what obstacles are most important, but the issues might include development of new reference standards for these replacement tools, examination of the response of these tools relative to the americium-beryllium tools, and exploration of any differences in response when the replacement tools are used in combination with other nuclear and nonnuclear well logging tools.

TABLE S-2 Summary of Radionuclide Radiation Source Uses and Possible Replacements

	(a) Radiotherapy (b) Gamma Knife®	Panoramic Irradiators	Prevention of GVHD (Blood Irradiation)	Radiography	Well Logging Porosity Measurement
Radionuclide Radiation Sources					
Radionuclide and activity	(a) Co-60: 500 TBq (15 kCi) (b) Co-60: 220 TBq (6 kCi)	Co-60: 100,000 TBq (3 MCi)	Cs-137: 40–100 TBq (1–3 kCi)	(a) Ir-192: 4 TBq (100 Ci) (b) Co-60: 0.25–4 TBq (5–300 Ci)	Am-Be: 0.25–0.8 TBq (8–22 Ci)
Primary device suppliers	(a) None currently sold in the United States (b) Elekta	MDS Nordion	MDS Nordion J. L. Shepherd CIS	SPEC, QSA Global, and others	Schlumberger, Baker-Hughes, Halliburton, and others
Capital cost	(a) None currently sold in the United States (b) $4M for machine, $2M for bunker	Approximately $54 per TBq ($2 per Ci), or $6M–10M for a large facility	$150,000–$225,000	(a) $8,000 for system $0.4/GBq for source (b) $30,000 for system $4/GBq for source	$30,000–$80,000 for the source, depending on activity and encapsulation requirements
Lifetime	Reload about every 5 years	Annual partial reload	30 years	(a) 3 months, (b) 5 years	Decades
Radiation Source Possible Replacements					
Replacement technology	Dedicated or specialized radiotherapy linac	Electron accelerator to make electron beam or x-ray beam	X-ray irradiation (a) Tubes (b) Linacs (c) Cs-137 robust forms (d) Co-60 (e) Filtration (f) Chemical treatment	(a) Pulsed x-ray (b) Compact accelerator (c) Phased-array ultrasonics	(a) D-T (fusion accelerator source) or (b) Californium-252 source
Replacement device suppliers	(a) Elecktra, Siemens, Varian, and others (b) Accuray, BrainLab	IBA Varian Others could be interested	(a) MDS Nordion (b) Many (c) Not now available (d) Not currently sold (e) Not yet approved (f) Not yet approved	Many	(a) Schlumberger (b) Pathfinder
Capital cost	(b) $4M for machine, $2M for bunker, unless linac is shared for standard radiotherapy	$10M for a large x-ray facility	(a) $150,000 (b) $3M or $0 (if already in house)	(a) Approx. $50,000 or more (b) Approx. $200,000 (c) Ranges $50,000–$100,000	(a) Estimated at $40,000–$50,000 based on other D-T sources (b) $5,000–$6,000, replace more often than Am-Be
Operating cost compared to radionuclide option	Lower than Gamma Knife®	Somewhat higher than gamma irradiator	(a) Higher than gamma irradiation (b) Higher than gamma irradiation	(a) Higher than gamma radiography (b) Higher than gamma radiography (c) Higher, technician requires more training	Similar to radionuclide source
Lifetime	10–15 years (obsolescence of computer controls)	Perhaps comparable to gamma irradiator		Unknown	(a) 4–5 years (b) 4–5 years
Comments	Some doctors prefer Gamma Knife® over linac options	Viability not yet proven against contract gamma irradiators	(a) Currently only one model available (b) Backup option only	Some applications still require radionuclide radiography	Commercially available today, but not yet widely adopted.

CHAPTER 1

INTRODUCTION

On September 28, 2006, a man identifying himself as Abu Hamza al-Muhajir, thought to be the leader of al-Qaeda in Iraq, appealed in an internet recording to experts in the sciences, "especially nuclear scientists and explosives experts" to join the field of jihad or holy war by using unconventional weapons, including dirty bombs, on American targets (Rising, 2006). Since the attacks on the United States on September 11, 2001, such threats are taken much more seriously. A recent report on preventing radiological terrorism states that "The likelihood of stolen Russian [ionizing radiation sources] being smuggled into the United States seems relatively low since a terrorist group would probably try to obtain a [radiation source] that is already located in the United States rather than risk detection at a point of entry into the country" (National Research Council, 2007). Whether terrorists would be able to obtain or gain access to a significant quantity of radioactive material and carry out an attack is a matter for analysis and debate, but the availability of high-intensity radiation sources is an important element of the risk (see Chapter 3 for a more detailed discussion of risk).[1]

Radiation sources come in many different types, forms, and intensities. The International Atomic Energy Agency (IAEA) issued a revised categorization system for radiation sources in 2003, ranking them according to the hazards they pose in descending order from Category 1 to Category 5 (IAEA, 2005a). (The definition of each category is provided in this chapter.) According to a survey by the U.S. Nuclear Regulatory Commission (U.S. NRC, 2007a), as of 2006 there were approximately 28,200 civilian Category 1 radiation sources in approximately 1,000 devices[2] in the United States, sources that, if not safely managed or securely protected, would be likely to cause permanent injury to a person who came in contact with them for more than a few minutes, and would cause fatal exposures in a few minutes to an hour, if not shielded. Another roughly 25,500 sources in approximately 4,000 devices could cause permanent injury to someone in contact with them for a short time (minutes to hours), and without shielding could be fatal to a person exposed for a period of hours to days (Category 2 sources).

Historically, the U.S. NRC and the Agreement States have issued and kept track of materials licenses (who is licensed to have radiation sources, what the sources are used for, and how much the licensee is permitted to hold). In the period immediately following the September 11 attacks, the U.S. NRC was unable to tell decision makers in the administration and Congress the number and locations of radiation sources in the United States. This was because neither the U.S. NRC nor most of the Agreement States maintained their own inventories of the actual radiation sources held by licensees and the locations of those sources, although the licensees themselves were required to maintain records on their own sources. The U.S. NRC took steps to remedy that situation, conducting voluntary surveys in 2004, 2005, and 2006 to learn the size and scope of the challenge entailed by source tracking and to establish

[1] The terms radiation source and sealed source refer to encapsulated radioactive material. Devices such as x-ray tubes and linear accelerators can generate radiation and are referred to in this report as radiation generators, when not specifically identified by name.

[2] The distinction between devices and sources is important. Many sources may be contained in a single device. For example, a Gamma Knife® used for treatment of brain lesions contains 201 cobalt-60 sources, each of which is a Category 2 source. The loaded device is a Category 1 device because of the aggregation of sources. This study examines possible replacements of radiation sources based on their applications, so it is focused mostly on devices, looking at replacing all of the radiation sources in a device or replacing the device with one that does not use radioactive material.

communication channels. Each year, the U.S. NRC has improved its reporting and identified additional licensees that should be surveyed, and each new survey has accounted for several thousand sources not included in the previous year's tally. A comprehensive radiation source tracking system for all Category 1 and 2 sources is to be in place by the fall of 2008 (U.S. NRC, 2007b).[3]

Concern about the potential for a radiological attack, the lack of information about the quantities and character of radiation sources in civilian use, and the modest level of security evidently afforded to many radiation sources at the time prompted Representative Edward J. Markey of Massachusetts and Senator Hillary Rodham Clinton of New York to sponsor a bill called the Dirty Bomb Prevention Act. The bill was not enacted into law, but language from the bill was included in the 2005 Energy Policy Act (commonly abbreviated as EPAct). A section of the EPAct is devoted to security of radiation sources, defined as high-hazard radioactive material. The section requires several actions by the federal government, including requirements that the U.S. NRC establish the national source tracking system mentioned earlier, lead an interagency task force to report to Congress on reducing the risk of radiological attacks (often referred to as dirty bomb attacks or radiological dispersal device [RDD] attacks), and request this study by the National Research Council of the National Academies on radiation source use and replacement. This study identifies the uses of radiation sources and how feasible it is to reduce the hazard from these sources by finding alternative means to accomplish their tasks, or by using less hazardous radioactive sources (see the full statement of task in Sidebar 1-1).

The interagency task force issued its first report to Congress on radionuclide radiation source protection and security (U.S. NRC, 2006a), and the Department of Energy issued its report on alternatives to industrial radioactive sources in 2006 (DOE, 2006). There is a growing body of research and reports on radiological security and consequence management related to incidents with radionuclide radiation sources (see, e.g., Harper et al., 2007; IAEA, 2006a, 2003a,b, 2001; Musolino and Harper, 2006; CRCPD, 2006; ANL, 2005; Ferguson and Potter, 2004; Medalia, 2004; Zimmerman and Loeb, 2004; CDC, 2003; DHS, 2003; DOE/U.S. NRC, 2003; Ferguson and Lubenau, 2003; Ferguson et al., 2003; GAO, 2003; Van Tuyle et al., 2003; NCRP, 2001). The U.S. NRC and Agreement States develop, implement, and enforce regulations that protect against radiation exposures, both accidental and malicious. In addition, radiological safety is the central mission of the radiation protection systems recommended by the International Commission on Radiological Protection and the National Council on Radiation Protection and Measurements. Consideration of technological alternatives to radionuclide radiation sources also has been part of the radiation protection efforts (for an overview see Lubenau and Strom, 2002), and is recommended by the Health Physics Society (2006). This study builds on the work cited here to address the statement of task.

SCOPE OF THE STUDY AND OVERVIEW OF THE REPORT STRUCTURE

This study is not meant to be a review of all radiation sources. The U.S. NRC study request explicitly confines the study to Category 1 and 2 sources. Category 3 sources are mentioned as illustrations of particular points in the report (see, e.g., gamma-ray tools in Chapter 9) and as possible future concerns to the U.S. NRC. Further, both the U.S. NRC and Congress are chiefly interested in radiation source uses and possible replacements in the United States. Some radiation sources have uses outside the United States. These have only been included in the report if there is a reasonable expectation that the use might be adopted in the United States or to provide context.

[3] The U.S. NRC has directed its staff to evaluate extending reporting and additional security requirements to Category 3 sources. The IAEA system suggests that a sliding scale of additional security measures be applied according to category.

SIDEBAR 1-1
STATEMENT OF TASK

The principal task of this study is to review the current industrial, research, and commercial (including medical) uses of radiation sources to identify uses for which:

1. the radiation source can be replaced with an equivalent (or improved) process that does not require the use of radioisotopes; or
2. the radiation source can be replaced with another radiation source that poses a lower risk to public health and safety if it is involved in an accident or used in a terrorist attack.

The study should explicitly consider technical and economic feasibility and risks to workers from such replacements.

The National Academies will issue a public report at the conclusion of the study. The report will contain a review of radiation source use, potential replacements for sources that pose a high risk to public health or safety, and findings and recommendations on options for implementing the identified replacements.

The committee has focused primarily on identifying the radiation source applications and evaluating the technological options for replacing the radiation sources currently used in those applications. Evaluating the radiation source security and risks, and more specifically the probabilities and consequences of a radiological attack using radiation sources, is outside the committee's charge. The committee does, however, consider these factors in general terms because they help to prioritize among the radiation sources that should be considered for replacement and because they affect costs and perceived costs of using radiation sources.

The statement of task states that the report should "explicitly consider technical and economic feasibility." The committee interprets "economic feasibility" to mean that a possible replacement should not be economically prohibitive, not that a replacement may be more (or less) expensive if it would provide the similar technical outputs. This element of the task statement does not require cost-benefit analyses, although such analyses may be valuable and necessary as part of implementing replacements (see Chapter 10).

DEVELOPMENT OF A RADIATION SOURCE CATEGORIZATION SYSTEM

Many of the radiation sources used throughout the world in medicine, industry, agriculture, research, and education are in the form of sealed sources, which are radioactive materials contained or bound in a solid material and encapsulated, typically in one or two welded stainless steel containers. The many applications of sealed sources, involving a variety of radioactive materials in a wide range of quantities, require varying levels of control and security according to the hazards the sources pose. Because they are sealed, intact sources typically present a risk of external radiation exposure only; that is, they irradiate tissue from outside the body. Sources that are leaking or have been punctured or broken can also cause internal radiation exposure—irradiation from inside the body, if ingested or inhaled—and contamination of the environment. However, as is described in Chapter 3, economic and social effects may be the most significant consequences of malicious use of radioactive material.

The safety and security mechanisms in place for radioactive sources vary widely from country to country based in part on the differences in regulatory infrastructure for controlling radiation safety and security. This has led to concern among national and international agencies responsible for the safe use and transport of radiation sources.

Building on efforts begun in the 1990s and motivated by the September 11 attacks, the IAEA[4] Board of Governors approved a revised Code of Conduct on the Safety and Security of Radioactive Sources in 2003 (IAEA, 2004a) using a system of source categories described in Categorization of Radioactive Sources (IAEA, 2005a). The Code of Conduct provides guidelines for countries in the development and harmonization of policies, laws, and regulations on the safety and security of radioactive sources. The categorization system was developed to provide for a logical international risk-based ranking of radioactive sources based on their potential for harm to human health and for grouping of source use practices into discrete categories. It was also intended to address the need for governments and regulatory authorities to make risk-informed decisions in establishing regulatory infrastructures, improving control over radioactive sources (including regulatory measures, registries, and import/export controls), prioritization of regulatory resources, preparing for and responding to emergencies, optimizing security measures for radioactive sources, including potential malicious use, and addressing other issues.

Basis for the IAEA Categorization System

Radiation sources pose low risks to radiation workers and the public when they are managed safely and securely. When mismanaged, however, they can cause an array of deterministic effects[5] that can lead to acute radiation sickness, permanent damage to limbs and organs, and death, depending on the source and how it is mishandled. The IAEA chose human health and safety as the primary attribute of importance in the development of the categorization system, and focused on the potential to cause deterministic health effects as the basis for the system.

The potential for harm involves not only the physical properties of the source (the radionuclide, type of radiation emission, and activity of the radiation source) but also the way in which the source is used. The actual practice in which the source is used and the shielding provided by devices containing the sources were also considered. Factors that were not considered include socioeconomic consequences of accidents or malicious acts, stochastic effects, such as the increased risk of cancer, and use of sources for medical reasons.

The structure of the categorization system is based on a threshold level of risk associated with deterministic effects, above which a source is considered "dangerous"[6] because it could cause a fatal or life-threatening exposure or result in a permanent injury that decreases the quality of life. The IAEA threshold dose levels corresponding to these risks differ based on

---

[4] One of the primary missions of the IAEA is to help countries upgrade nuclear safety and security, and prepare for and respond to nuclear and radiological emergencies. To fulfill this mission, IAEA plays an instrumental role in developing international conventions, standards, and expert guidance, with input from the agency's member countries. The IAEA focuses its efforts regarding radiation sources on assisting countries to protect people and the environment from harmful radiation exposure.

[5] Effects associated with radiation are described as either deterministic or stochastic. Deterministic effects manifest themselves in a relatively short time after a high-intensity exposure to radiation (e.g., 1 or more sieverts) and can range from erythema (skin redness) to disruption of body-organ functions. Stochastic effects are increased risks of various maladies that manifest themselves over a longer time period. Most notable among the stochastic effects is induction of cancer, but heart disease and other conditions can also result, depending on the exposures.

[6] As defined in the IAEA Safety Standards Series No. GS-R-2, Preparedness and Response for Nuclear or Radiological Emergency, a "dangerous source … [is] a source that could, if not under control, give rise to exposure sufficient to cause severe deterministic effects." A severe deterministic effect is one that is fatal or life-threatening or results in a permanent injury that decreases the quality of life. The dose considerations included not only external exposures to a bare (unshielded) source, but also dispersal of a source, for example, by fire, explosion, or human action, resulting in a dose from inhalation, ingestion, or skin contamination. See IAEA (2005a).

the organ affected, such as 1 Gy delivered to the whole body (see Sidebar 1-2),[7] 1 Gy delivered to the bone marrow in 2 days, 6 Gy delivered to the lung in 2 days, 5 Gy delivered to the thyroid in 2 days, or 25 Gy absorbed in skin or surface tissue at a depth of 2 cm for most parts of body or 1 cm for the hand for a period of 10 hours (IAEA, 2005a, 2006).

The authors of the system calculated the activity, referred to as the 'D' value, of each relevant radionuclide corresponding to that threshold risk level. This provides a relative ranking of radioactive sources and the practices in which they are used. Devices, including single-source devices, are classified into five categories, according to their potential for causing harmful health effects if not managed safely and securely. There is flexibility in the system, in that, although common practices (such as high-dose-rate brachytherapy) are grouped in one category, particular sources may be assigned to a category based on their activity (A) alone, by dividing their activity by the D value, resulting in an A/D ratio. Also, aggregations of sources in one location can be categorized by summing their A/D ratios. Although the categorization system was focused on sealed sources, it may also be applied to unsealed radioactive material in some situations.

The definitions of the five categories, provided in plain language in IAEA (2005a), are listed below, and Table 1-1 lists the A/D ratios and examples of practices for each of the five categories in the categorization system. If not managed safely or securely,

- Category 1 sources could lead to the death or permanent injury of individuals in a short period of time.
- Category 2 sources could lead to the death or permanent injury of individuals who may be in close proximity to the radioactive source for a longer period of time than for Category 1 sources.
- Category 3 sources could lead to the permanent injury of individuals who may be in close proximity to the source for a longer period of time than Category 2 sources. Sources in Category 3 sources could but are unlikely to lead to fatalities.
- Category 4 sources could lead to the temporary injury of individuals who may be in close proximity to the source for a longer period of time than Category 3 sources.
- Category 5 sources could cause minor temporary injury of individuals, but are unlikely to do so.

---

[7] Throughout this report, quantities are reported in SI units, which are explained in Sidebar 1-2. A glossary of terms can be found in Appendix C.

SIDEBAR 1-2
RADIATION QUANTITIES AND UNITS

The metric system of units known as the Système International d'Unités (International System of Units, or SI units) is based on units for seven basic physical quantities; all other quantities and units are derived from the basic quantities and units. The basic physical quantities are listed in the first table below. The radiation and radiological units cited in this report are derived units, defined in the second table.

SI Base Units

Name	Symbol	Quantity
Meter	m	Length
Kilogram	kg	Mass
Second	s	Time
Ampere	A	Electric current
Kelvin	K	Thermodynamic temperature
Mole	mol	Amount of substance
Candela	cd	Luminous intensity

SI Derived Units Relevant to This Report

	Name	Symbol	Expressed in Other SI Units	Expressed in SI Base Units
Energy, amount of heat	Joule	J	N m	$m^2\ kg\ s^{-2}$
Power, radiant flux	Watt	W	J/s	$m^2\ kg\ s^{-3}$
Electric charge	Coulomb	C		s A
Electric potential difference	Volt	V	W/A	$m^2\ kg\ s^{-3}\ A^{-1}$
Celsius temperature	degree Celsius	°C		K
Activity[a]	becquerel	Bq		s^{-1}
Specific activity	becquerel per gram	Bq/g		$s^{-1}\ g^{-1}$
Absorbed dose[b]	Gray	Gy	J/kg	$m^2\ s^{-2}$
Dose equivalent[c]	Sievert	Sv	J/kg	$m^2\ s^{-2}$
Exposure (x- and γ -rays)[d]	Coulomb per kilogram	C/kg		$kg^{-1} s\ A$

[a] *Activity* of a radioactive substance is defined as the number of decays per time. Its SI unit is the becquerel (Bq) corresponding to one radioactive decay (disintegration) per second; its old unit, the curie (Ci), was originally defined as the activity of 1 gram of radium-226 and later as 3.7×10^{10} Bq.
[b] *Absorbed dose* or *dose* is defined as the energy absorbed per unit mass of medium. Its SI unit, gray (Gy), is defined as 1 joule (J) of energy absorbed per kilogram of absorbing medium; its old unit is the rad, defined as 100 erg of energy absorbed per gram (g) of absorbing medium, which is 0.01 Gy.
[c] *Dose equivalent* is defined as the dose multiplied by a radiation-weighting factor to account for the differences in biological harm to human organs that result from differences in radiation type and energy for the same physical dose received by the organ. The SI unit of equivalent dose is the sievert (Sv); the old unit is the rem, which is equal to 0.01 Sv. For x rays, gamma rays (γ rays), and electrons the weighting factor is 1; for protons it is 5; for alpha particles it is 20; and for neutrons it ranges from 5 to 20 depending on neutron energy.
[d] *Exposure* is related to the ability of photons to ionize air. Its old unit, roentgen (R), is defined as charge of 2.58×10^{-4} C produced per kilogram of air.
SOURCE: Adapted from Tables 3 and 4 from International System of Units (SI) (2006).

TABLE 1-1 IAEA Radiation Source Categories

Category	Activity Ratio	Examples of Practices and Devices	Examples of Threshold Activity Levels (TBq)	
1	A/D > 1,000	Radioisotope thermoelectric generators (RTGs), panoramic irradiators, large self-shielded irradiators, teletherapy, fixed multibeam teletherapy (Gamma Knife®)	americium-241 cobalt-60 cesium-137 iridium-192	60 30 100 80
2	1,000 > A/D > 10	Smaller self-shielded irradiators, industrial gamma radiography, well logging devices	americium-241 californium-252 cobalt-60 cesium-137 iridium-192	0.6 0.2 0.3 1.0 0.8
3	10 > A/D > 1	High- and medium-dose-rate brachytherapy, fixed industrial gauges (level gauges, dredger gauges, high-activity conveyor gauges, spinning pipe gauges), well logging devices	americium-241 cobalt-60 cesium-137 iridium-192	0.06 0.03 0.1 0.08
4	1 > A/D > 0.01	Low-dose-rate brachytherapy (except strontium-90 eye plaques and implant sources), thickness gauges, portable gauges, bone densitometers		
5	0.01 > A/D > Exempt quantity/D	X-ray fluorescence devices, static eliminators, electron-capture devices		

NOTE: 1 TBq = 27 Ci. SOURCE: Adapted from Table 1 of IAEA (2004a).

PREVIEW OF TRADE-OFFS AND METHODOLOGICAL APPROACH

Radiation sources have many valuable uses in medicine and industry. They contribute to the production of such final goods as sterile medical equipment, blood products that do not promote graft versus host disease in transplant recipients, cancer treatment, oil and gas development, and product quality assurance. A business or other organization chooses to use radiation sources over other alternatives because it perceives the benefits from using the sources to be greater than the costs it bears to use them. The perceived benefits arise because a business believes that the radiation sources allow it either to accomplish tasks that may not be feasible with other available technologies or to accomplish these tasks less expensively than with other technologies. The costs borne by the business include both direct financial costs of use (capital and operating) as well as indirect financial costs related to the safe and legal requirements of use (security, insurance, regulatory compliance, safety training). If all of the costs borne by society were included in the costs seen by radiation source users, then an efficient, and one could argue, socially desirable pattern of use would result. However, if the costs borne by society are substantially higher than those borne by users, then an inefficient and socially undesirable pattern of use may result. The commissioning of the study was motivated primarily by concern about one potentially large social cost: diversion of radiation sources for use in terrorism.

Examining the study task in terms of costs applies a useful structure to the enquiry, pushing for quantification of costs where possible, and describing costs where they cannot readily be quantified. For the purposes of this study, the total social costs (TSC) of use of a radiation source technology can be divided into three components: the private costs of use (PUC), the costs associated with terrorism risks (TRC), and other social costs (OSC) for costs not borne by the user aside from those related to terrorism:

$$TSC \equiv PUC + TRC + OSC \qquad (1\text{-}1).$$

In common economics terminology, PUC is the internal cost of use, and the sum of TRC and OSC is the external cost of use. As explained in Chapter 3, TRC primarily reflects the increased risk of the use of an RDD that would be borne by those persons exposed, the owners of the affected property, and the governments responsible for responding and mitigating the effects. OSC reflects other external costs, such as those related to safe disposal of spent sources or risks to employees not fully borne by firms through health and safety liability.

If it were possible to monetize the external costs confidently, then one could imagine imposing user fees or regulatory burdens such that the private costs seen by users (PUC) equaled the TSC. Internalizing the external cost would create incentives for users to seek out the most desirable technology from the social perspective, perhaps even abandoning some uses altogether. The responses of users to the TSCs in turn would induce suppliers of radiation sources to search for alternative technologies with lower total social costs so that efficiency would increase over time.

Implementing such an approach, however, is not practical because the external costs cannot be confidently monetized. Most importantly, it is not possible to monetize the terrorism risks (TRC) because we do not have a firm basis for predicting the relationship between particular radiation source uses and the expected costs of terrorism. While it may be possible to identify representative scenarios of RDD deployment or other acts of terrorism involving radiation sources, it is not possible to quantify the probabilities of these scenarios or how any particular type of radiation source contributes to them. Consequently, the committee cannot recommend a simple algorithm in seeking to identify desirable radiation source replacements.

The methodological approach of the committee can be described in terms of a number of discrete steps. However, these steps were not and could not be followed in strict order. Rather, they convey the overall logic of the committee's efforts.

Step 1: Identify radiation source technologies with relatively large TSCs

The divergence between private and social costs of the use of radiation source technologies is primarily a function of their contribution to risk of the use of radiation sources in a terrorist attack and the actual cost of disposal of the used source, for sources that cannot now be disposed. The size of TSC is a function of a number of factors: the quantity of radioactive material used in the technology, the ease with which the material can be diverted from the technology, the likelihood of timely discovery of the diversion, the ease with which the material can be dispersed, and the costs of responding to a dispersion (including both the cost of cleanup and the consequences of the exposures).

Step 2: Identify alternative technologies offering potentially large reductions in TSCs

Technologies with large TSCs offer the greatest potential for overall reductions in the risk of radiological terrorism. There are five general types of alternative technologies to consider.

1. It may be possible to replace the radionuclide radiation source–based technology with a technology not involving radiation. For example, ultrasonics technology can substitute for some types of radiography.
2. An alternative technology may replace the use of radioactive sources with x-rays or an electron beam from an x-ray tube or electron accelerator. Similarly, americium-beryllium sources may be replaced by neutron generators (particle accelerators that generate neutrons from targets).

3. An alternative technology may replace one radionuclide with another that poses less risk. For example, americium-241 beryllium sources, which are of concern in part because of americium-241's long half-life, might be replaced in some applications by californium-252, which has a much shorter half-life and higher specific activity (requiring a smaller quantity overall).
4. The chemical and mechanical form of the radiation source may be changed to make it less valuable in terrorist diversion. For example, cesium-137 might be incorporated into a mineral (pollucite) or polycrystalline ceramic to make it harder to disperse than the more common form, cesium chloride.
5. Looking over the entire life cycle of use, alternative approaches to security, transportation, and disposal may be introduced (see, e.g., Van Tuyle et al., 2003).

The committee searched for technically feasible alternatives that offer reductions in TSC, ΔTSC. The delta symbol, Δ, is used here to indicate change, so ΔTSC is a reduction or increase in TSC.

Step 3: Assess the implications for changes in private costs, ΔPUCs, of alternatives

Technically feasible alternatives may involve substantial increases in the private costs of use (ΔPUC). Alternative technologies that are currently available commercially, but are not widely used, will generally involve higher overall private costs than the technology in use or else they would already be in use. The higher costs may involve more costly capital, higher operating costs, or costly adjustments in other aspects of operations, such as irradiation of smaller batch sizes or multiple measurements with less effective measuring devices. When the magnitudes of both ΔTSC and ΔPUC are large, it may be socially worthwhile to consider public policies to make the technological transition more financially attractive or at least feasible for users.

Step 4: Assess the implications for changes in other social costs, ΔOSC, of alternatives

Alternative technologies may increase or decrease the other social costs involved in current use. For example, replacing radioactive sources with radiation generators may reduce disposal costs that are not currently borne by users (those that are not part of PUC). Moving from one radionuclide to another might change the exposure risks to employees or the transportation risks. It is also important to consider the fate of capital equipment and radiation sources used in the replaced technology.

Viewing these steps comprehensively, the approach of the committee can be thought of as seeking alternatives to current uses for which the difference between all of the costs of the radiation sources and the alternative technologies has the largest negative value:

$$\Delta TSC \equiv \Delta PUC + \Delta TRC + \Delta OSC \qquad (1\text{-}2).$$

That is, the committee sought to identify opportunities where alternative technologies appear to offer reductions in TSC.

It is worth noting several limitations of this methodological approach. First, the committee tried to use actual costs wherever possible, although some of these costs are approximate at best, and some costs simply could not be assessed. Second, the committee does not have sufficient resources or time to look far beyond existing technology. In particular, the costs of realizing alternative technologies that are physically feasible but not yet commercially available can only be assessed roughly. Finally, external components of TSCs can also only be assessed qualitatively. Still, even with these limitations, the committee chose this approach because its structure makes explicit the key considerations in evaluating implementation of alternatives, and enables the committee to discuss costs, even if they are not readily expressed in terms of dollars.

CHAPTER 2

RADIATION SOURCES IN THE UNITED STATES AND THEIR USES AND ORIGINS

SUMMARY

As a first step in understanding the uses of radiation sources in the United States, and to lay the groundwork for finding possible replacements, the committee examined the physical, chemical, and radiation characteristics of the radiation sources in use or available for manufacture; who uses the sources; and how they are applied. To carry out this aspect of its charge, the committee examined information collected by the U.S. Nuclear Regulatory Commission (U.S. NRC) and the U.S. Department of Energy (DOE), solicited input from representatives of the source manufacturers and distributors and from end users in several sectors, and carried out background scientific literature surveys. This chapter presents the results of those investigations. Each of the major applications is described in greater detail with an examination of possible replacements in Chapters 5 through 9.

RADIONUCLIDES: THEIR FORMS IN CATEGORY 1 AND 2 RADIATION SOURCES AND THEIR DISPOSITION PATHS

Radionuclides are types of unstable atomic nuclei. Each radionuclide can be identified by its chemical element (e.g., cesium), atomic mass (different isotopes of an element have different masses, e.g., cesium-135 and cesium-137), and its energy state in the case of metastable radionuclides (barium-137m). Critical characteristics of a radionuclide are its half-life, its mode of decay (alpha [α], beta [β], electron capture, spontaneous fission, and gamma emission [γ]), and the energies associated with any radiation emitted. For radionuclides whose decay products are also radioactive, one must also consider the same characteristics of the decay product. For this study, the chemical form and structure of the bulk that incorporates the radionuclide (e.g., cesium chloride in granular salt form) is also important. Before examining each important radionuclide and the major radiation source devices, it is useful to understand the disposition or disposal paths available for these sources.

The waste classification system for disposal of civilian low level radioactive waste in the United States is based on the harm the waste might cause 100 to 500 years in the future (see Title 10 of the *Code of Federal Regulations*, Part 61). U.S. NRC-regulated low-level waste is categorized as Class A, B, C, or Greater-than-Class C (in order of ascending hazard) based on the concentration of the radionuclide within its waste form.[1] Classes A, B, and C waste can be disposed in near surface disposal facilities, such as the Barnwell Waste Management Facility in South Carolina and the US Ecology facility near Richland, Washington.[2] The federal

---

[1] The concentration in the waste form may be quite different from the concentration when in use because of radioactive decay and because of packaging: In some cases, the concentration of the waste form for a source the size of a thimble may be averaged over a 200-liter drum in which it is packaged for transportation and disposal. See, e.g., GAO (2005).

[2] South Carolina law will close the Barnwell facility to waste from all states except New Jersey, Connecticut, and South Carolina after June 30, 2008. The Richland facility is open only to Colorado, Idaho, Montana, New Mexico, Nevada, Alaska, Hawaii, Oregon, Utah, Washington, and Wyoming.

government (DOE) is responsible for the disposal of Greater-than-Class-C waste (P.L. 99-240), which the U.S. NRC has determined is not generally suitable for near-surface disposal. DOE is in the process of planning to develop an environmental impact statement on options for Greater-than-Class-C waste,[3] but no civilian disposal facility is now available for this type of waste. Relatively short-lived radionuclides, such as cobalt-60, have no upper concentration limit for Class C (or even Class B, in the case of cobalt-60) because the radioactive material decays sufficiently over the centuries for the hazard to diminish below regulatory limits. As a practical matter for disposal, however, there are also waste acceptance criteria at the disposal sites and these criteria preclude some high-activity sources because of worker exposure limits.

DOE has disposal facilities for and regulatory authority over its own radioactive waste. DOE facilities operate under the same safety requirements (i.e., dose rate limits to workers, the public, and inadvertent intruders), but they do not use the Class A, B, and C system and can dispose of radiation sources with fewer impediments. Those facilities, however, only accept DOE waste.

The National Nuclear Security Administration's Offsite Source Recovery Project (OSRP) is responsible for recovery of sealed sources that represent threats to public health and safety and security. The OSRP recovers unwanted and abandoned sources, particularly sources that have no disposal path available.[4] The OSRP maintains or contracts for short- or long-term storage, recycles or reuses radioactive material when appropriate, and disposes of recovered sources if an appropriate disposal site is available (DOE or commercial; U.S. NRC, 2006a). Most of the recovered sources are stored onsite at Los Alamos National Laboratory or at a contractor's facility. The OSRP will take and store the sources free of charge, so licensees are not stuck with their sources forever, and abandoned sources and sources held by bankrupt licensees are secured. A licensee can pay for delivery of its sources to the OSRP so that it can rid itself of them on its own schedule rather than on the OSRP's schedule. The OSRP is discussed further in Chapters 9 and 10.

The Conference of Radiation Control Program Directors (CRCPD) has a program partially funded by DOE for assisting licensees and agencies that are in possession of unwanted radioactive material.[5] The CRCPD "offers assistance in finding affordable, legal disposition for radioactive material through: Storage for decay, adoption by an individual, reuse by a device manufacturer, reprocessing of the material, acceptance by state or federal government, and commercial storage" (see CRCPD, 2007). The program provides education, guidance, and assistance in arranging for appropriate disposition of the radionuclide radiation sources, and funds for state and local governments to dispose of the radioactive material. One of these disposition options is to connect licensed parties seeking radiation sources with those who possess unwanted sources. (See U.S. NRC, 2006a, for a full description of options for

---

[3] DOE recently issued a Notice of Intent to Prepare an Environmental Impact Statement for the Disposal of Greater-Than-Class-C Low Level Radioactive Waste (*Federal Register*, 72, No. 140, (July 23, 2007):40135).

[4] The OSRP is not the first program for recovering unwanted radionuclide radiation sources. When cesium-137 brachytherapy sources began to replace radium sources (and to encourage the switch) the U.S. Public Health Service funded a program to recover and dispose of radium sources that were disused, unwanted, or orphaned. The Environmental Protection Agency (EPA) took over the program, which ended in 1983. Subsequently, the Conference of Radiation Control Program Directors (CRCPD), with EPA support, mounted another radium recovery program. Together, these programs recovered 440 grams of radium (Lubenau, 1999). The CRCPD continues to assist states in retrieving and disposing of radioactive sources through its Orphan Sources Initiative. "In certain limited cases, the EPA and DOE, through CRCPD, provide funds to state radiation control programs for the disposition of radioactive sources when the owner cannot afford the costs of disposition or should not be held liable for those costs" (National Research Council, 2006a).

[5] A member of the authoring committee for this report is executive director of the CRCPD.

disposal of radiation sources.) The radionuclides in Category 1 and 2 sources in the United States are summarized in Table 2-1 and described in some detail along with their production and disposition options in the text that follows.

TABLE 2-1 Summary of Radionuclides in Category 1 and 2 Radiation Sources in the United States[a]

Radionuclide	Half-life	Radioactive Emissions and Energies	Typical Specific Activity (TBq/g) [Ci/g]	Total Activity in U.S. Inventory (TBq) [Ci]	Major Applications	Typical Activity (TBq) [Ci]	Physical or Chemical Form
Americium-241	432.2 yr	α–5.64 MeV, γ-60 keV, principal	0.13 [3.5]	240 [6,482]	Well logging	0.5–0.8 [13–22]	Pressed powder (americium oxide)
Californium-252	2.645 yr	α–6.22 MeV, fission fragments, neutrons, and gamma rays	20 [540]	0.26 [7]	Well logging	0.0004 [0.011]	Metal oxide
Cesium-137 (Ba-137m)	30.17 yr	β-518 keV max with γ-662 keV (94.4% of decays) or β-1.18 MeV max	0.75 [20]	104,100 [2.8 million]	Self-contained irradiators Teletherapy Calibrators	75 [2,000] 50 [1,400] 15 [400]	Pressed powder (cesium chloride)
Cobalt-60	5.27 yr	γ-1.173 and 1.333 MeV	3.7 [100] 11 [300]	7.32 million [198 million]	Panoramic irradiators Self-contained irradiators Teletherapy Industrial radiography	150,000 [4 million] 900 [24,000] 500 [14,000] 4 [100]	Metal slugs Metal pellets
Iridium-192	74 d	β-1.46 MeV max with 2.3 γ-380 keV average, 1.378 MeV max (0.04% of decays)	18.5 [500]	5,436 [146,922]	Industrial radiography	4 [100]	Metal
Plutonium-238	87.7 yr	α–5.59 MeV, and γ-43 keV (30% of decays)	2.6 [70]	34.7 [937]	RTG Pacemakers (obsolete) Fixed gauges	10 [270] 0.1 [3] 0.75 [20]	Metal oxide
Selenium-75	119.8 d	γ-280 keV average, 800 keV max	20–45 [530–1200]	9.7 [261]	Industrial radiography	3 [75]	Elemental or metal compound
Strontium-90 (Yttrium-90)	28.9 yr	β-546 keV	5.2 [140]	64,000 [1.73 million]	RTG	750 [20,000]	Metal oxide

[a] Nuclear decay data for this table and throughout the report are from Firestone and Shirley (1996).

Cobalt-60

Cobalt is a metal element with only one stable isotope: cobalt-59. When natural cobalt slugs are placed in a nuclear reactor, the nuclei absorb thermal neutrons to make cobalt-60, a radionuclide with a 5.27-year half-life. Cobalt-60 undergoes beta decay (emits an electron and a neutrino) and emits two gamma rays with each decay; one at 1.173 MeV and one at 1.333 MeV.[6] Cobalt-60 sources are produced as high-specific-activity sources for teletherapy and industrial radiography and industrial sources for irradiators and other applications. High-specific-activity sources are small pellets (typically cylinders 1 mm in diameter and height) of metal produced in specialized high-flux nuclear reactors (e.g., Atomic Energy of Canada Ltd.'s NRU reactor in Chalk River, Canada; the Advanced Test Reactor in Idaho[7]; and the Research Institute of Atomic Reactors' SM reactor in Dimitrovgrad, Russia), several thousand of which might be put into one source capsule to make a teletherapy source. Canadian CANDU power reactors produce the vast majority of the industrial cobalt-60 used in the United States (U.S. NRC, 2006a). The industrial targets are in the form of "pencils," which are sealed zircaloy tubes that house a stack of small, cylindrical cobalt slugs. A cobalt pencil is typically irradiated in the reactor core for approximately two or more years (Slack et al., 2003). Figure 2-1 shows a typical cobalt-59 pencil, slug, and pile of pellets before irradiation. The price of cobalt-60 sources varies based on the specific activity, total activity, and design of the source, ranging from about $40 to $53/TBq ($1.5 to $2/Ci) for industrial sources (see, e.g., Smith, 2006) to $215/TBq ($8/Ci) for a teletherapy source and $4,300/TBq ($160/Ci) for Gamma Knife® and radiography sources (S. Laflin, International Isotopes, Inc., personal communication with M. D. Lowenthal, July 19, 2007; G. Moran, Source Products Equipment Corporation, personal communication with M. D. Lowenthal, July 18, 2007).[8]

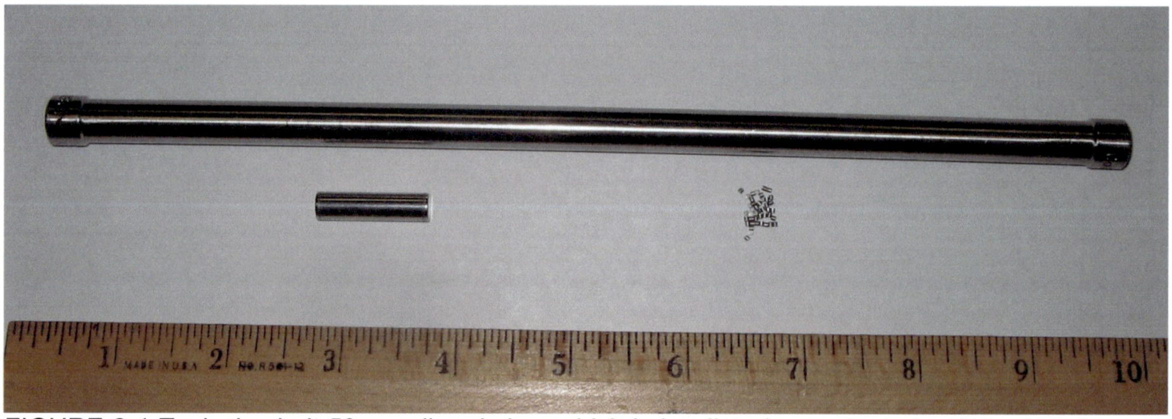

FIGURE 2-1 Typical cobalt-59 pencil and slug, which is irradiated to make industrial cobalt-60 sources. Also shown is a pile of cobalt-59 pellets, which can be irradiated in a high-flux reactor to make cobalt-60 teletherapy and industrial radiography sources. The ruler uses English units (1 inch = 2.54 cm). SOURCE: Image provided by committee.

---

[6] There are four other gamma rays from cobalt-60 with lower emission probabilities. Any of these may be emitted during a particular decay, but fewer than one in every 6,000 decays is accompanied by one of these rays, so they are of lesser importance.

[7] International Isotopes, Inc., contracts with DOE to have high-activity cobalt-60 produced in the Advanced Test Reactor.

[8] These costs, like the costs for radiation sources and devices cited throughout this report, reflect the figures quoted to the committee by manufacturers and or customers. They are not necessarily representative of all prices paid by customers, which vary based on the size of the order, special requests, and other business relationships or agreements.

Disposal of cobalt-60 sources in low-level waste disposal facilities is allowed under federal regulations; however, as noted above, high-activity sources are generally precluded by the waste acceptance criteria at the disposal sites because of worker exposure limits. The U.S. NRC (2006a) reports that the disposal costs at the Barnwell Disposal Facility in South Carolina are $1,870/ft^3$ of waste, plus surcharges including $11.32/GBq ($0.419/mCi), which totals about $130,000 for disposal of just one 11.1-TBq (300-Ci) cobalt-60 source.

In practice, high-activity cobalt sources typically are returned to the manufacturer and distributor or sometimes taken by the federal government under the OSRP. MDS Nordion and International Isotopes, Inc., for example, mix pellets of cobalt from moderately decayed sources with cobalt pellets fresh from the reactor in some newly fabricated sources to balance overall activity and achieve the specified activity level for a new source. A few other facilities will store cobalt-60 and other relatively short-lived radionuclides for decay. There is a cost associated with returning used cobalt-60 sources to a manufacturer and distributor. That cost varies according to the quantity and age of the material and the cost of transportation, but is typically in the tens of thousands of U.S. dollars.

The OSRP had recovered 606 cobalt-60 sources comprising 2,340 TBq (63,197 Ci) and registered another 442 sources totaling 16,759 TBq (452,481 Ci) as excess or unwanted, as of January 2007.

Cesium-137

Cesium is a highly reactive alkali metal element with one stable isotope: cesium-133. The radionuclide cesium-137, which is produced by fission in a nuclear reactor, has a 30.17-year half-life and decays by beta decay to barium-137, which is stable, in 15 percent of the decays and to become barium-137m, a metastable radionuclide, in 85 percent of the decays. Barium-137m decays to stable barium-137 with a half-life of 2.55 minutes, emitting a 661.7-keV gamma ray (see Figure B-2 in Appendix B). Radioactive cesium sources are mostly used in self-shielded irradiators, which take advantage of its moderate gamma energy (requiring moderate shield thicknesses) and its 30-year half-life (enabling a source to last for the lifetime of the device). Cesium-137 is produced by fissioning uranium nuclei and then chemically separating the cesium from the irradiated nuclear fuel or targets.[9] Most facilities that chemically process (reprocess) spent nuclear fuel to recover uranium and plutonium leave cesium in the waste stream. The cesium actually is made up of four isotopes: cesium-133 (stable), cesium-134 (2-year half-life), cesium-135 (2.3 million years), and cesium-137. Cesium-134 is produced in very small concentrations, which are reduced further by decay. Cesium-137 constitutes about 25 to 32 percent of the cesium atoms. All cesium atoms share in any cesium chemical reaction or compound so the cesium-137 concentration is diluted by the presence of other cesium isotopes.

Separated radioactive cesium sold internationally is produced only by the Production Association Mayak (PA Mayak), in the Chelyabinsk region of Russia and sold through the U.K.-based company, REVISS. It is supplied as cesium chloride, a crystalline salt (it is chemically and structurally related to table salt, sodium chloride) that can be made in a range of particle sizes, from centimeter-scale blocks to powder, as is used in the manufacture of radioactive cesium chloride sources. After cold-pressing to form a pellet inside a stainless steel thimble-shaped receptacle, the receptacle is loaded in a protective stainless steel capsule that is welded to form the inner containment, and a second stainless steel jacket is welded over the first to

[9] Because there are four atomic mass units between cesium-133 and cesium-137, and because xenon-136 (a material that would decay to cesium-137 after absorbing a neutron) is a noble gas that cannot be formed into a dense target, it is impractical to produce cesium-137 by irradiating low-atomic-number targets with neutrons (neutron activation).

form the actual sealed radioactive cesium chloride source. The production of radioactive cesium chloride sources is carried out at around 200°C because cesium chloride is hygroscopic.

Cesium chloride undergoes a change in crystal structure above 469°C with an accompanying 19 percent density reduction. It melts at 645°C with a further increase in volume (Zull, 1996). Because of the volumetric changes that occur on heating and cooling, which could distort or rupture the stainless steel container if the radioactive cesium chloride were packed into the capsule near its maximum density, the cesium chloride is emplaced in the capsule in the form of a porous, pressed pellet with a density of about 2.5 to 2.7 grams per cubic centimeter (g/cm^3). (The density of pure cesium chloride crystals is 3.99 g/cm^3.)

Cesium chloride is soluble in water at room temperature and so, if it is intentionally or accidentally removed from its container, it can readily be dispersed. If a leak in the stainless steel container were to occur, the cesium chloride could dissolve in water and contaminate the nearby environment, as happened in the water tank of a panoramic irradiator facility used for sterilizing medical devices in Decatur, Georgia, in 1988 when a radioactive cesium chloride source containment failed due to thermal cycling and stress corrosion cracking (Setser, 1990; see Chapter 3 for a description of this incident). Cesium chloride is highly reactive in the environment; binding to surfaces and even migrating into concrete (see Chapter 3). If it enters the body, it disperses wherever water goes and delivers a whole-body dose.

One approach to decreasing the problems posed by the very high solubility of cesium chloride in water is to use another compound containing cesium-137 as a direct replacement for the cesium chloride powder. Alternative, lower specific activity forms of cesium-137 sources are currently used for lower Category 3 sources used in some industrial process control and in well logging devices. Two vitrified forms (glasses) containing more dilute concentrations of cesium-137 than in radioactive cesium chloride have been used. One is prepared by absorption of the cesium-137 isotope into a silicate zeolite which is then heated to form a glass typically containing about 0.22 TBq/cm^3 (6 Ci/cm^3) of cesium-137, according to QSA Global (H. Evans, QSA Global, verbal presentation to the Committee on Radiation Source Use and Replacement, Irvine, California, February 1, 2007), a factor of 8 lower than radioactive cesium chloride, which contains 50 atomic percent (79 weight percent, which is 0.64 to 0.82 TBq/g [17 to 22 Ci/g] depending on isotopic mix or 1.7 to 2.1 TBq/cm^3 [45 to 58 Ci/cm^3] for a cesium chloride density of 2.6 g/cm^3). The other is formed by a sol-gel process from a cesium-137 isotope salt. In this process, the cesium is attached through ion exchange to solid particles of glass formative materials in a suspension. These particles are gelled through chemical reactions and then heated to form a glass pellet. According to QSA Global, this latter process typically produces glass with 0.65 TBq/cm^3 (18 Ci/cm^3) (H. Evans, February 1, 2007), which is a factor of 2.6 lower than radioactive cesium chloride. The glasses are substantially less soluble in water and consequently less easily dispersed than the cesium chloride pellets. Being brittle, they can, like cesium chloride, be dispersed in an explosion. The glass forms of cesium-137 sources rather than cesium chloride are used in well logging because there are currently well logging standards (10 CFR § 39.2) that stipulate that the form of the cesium source must withstand impact loading (forces applied suddenly) during service and resist long-term chemical attack in the event that they are lost belowground. Both of these glass processes are commonly used to make sources in the range from 0.0075 to 0.11 TBq (200 mCi to 3 Ci), although larger sources can be made, especially with a process line tailored to making larger sources.

In the past, another, lower specific activity form of cesium-137, but with higher cesium loading than the glass, was available. It is a compound commonly referred to by its mineral name, pollucite. Pollucite tumor irradiation sources were manufactured in France by ORIS-Bio International in the early 1990s but were apparently discontinued because the maximum specific activity attainable with pollucite is lower than that of cesium chloride. Pollucite, $(Cs,Na)_2.[Al_2Si_4O_{12}]$, has a density of 3.3 g/cm^3 and melts above 1900°C. Because it can contain up to 10 atomic percent cesium (43 weight percent, which is 0.35 to 0.45 TBq/g [9.5 to 12 Ci/g], or 1.2 to 1.5 TBq/cm^3 [31 to 40 Ci/cm^3]), its specific activity is lower by a factor of 1.5 than that of radioactive cesium chloride at a density of 2.6 g/cm^3. Pollucite is known to be substantially less soluble in water than cesium chloride; the naturally occurring mineral is the primary ore for cesium and requires a series of acid-base reactions to extract the cesium. Further studies have shown that pollucite resists attack by high-temperature water (Komareni and Roy, 1983) and under hydrothermal conditions (Minura et al., 1997). Indeed, based on its resistance to water attack, pollucite has been identified in several studies as the preferred host to tie up cesium-137 in the long-term immobilization of nuclear waste (Clarke, 1983; Komareni and Roy, 1983; Clarke et al., 1981).

Radioactive cesium pollucite is much more difficult to produce than radioactive cesium chloride, which makes the pollucite form not only more expensive but also more difficult to research. The main difficulty in producing pollucite, as well as other alternative compounds including the glass forms of cesium, is associated with the volatility of cesium at high temperatures. This volatility leads to the creation of more contamination and process waste than is created in production of the other material forms of radioactive cesium mentioned here. To minimize volatility requires low-temperature chemical processing together with the capture of any cesium-137 vapor. A number of chemical reactions have been demonstrated to produce pollucite powders, including a hydrothermal reaction from cesium hydroxide and solid aluminum metal and silica (MacLaren et al., 1999):

$$2\,CsOH\,(aq) + 2\,Al\,(s) + 4\,SiO_2\,(s) + 2\,H_2O\,(l) \rightarrow 2\,CsAlSi_2O_6\,(s) + 3\,H_2\,(g) \qquad (2\text{-}1).$$

Alternatively, cesium-137-substituted pollucite can be synthesized by mixing colloidal silica solution with solutions of aluminum and cesium nitrates in stoichiometric proportions, followed by evaporation and calcining. This is the method used by ORIS-Bio International, which also sintered the pellets to 1200°C (Hess et al., 2000). A cesium-137-substituted pollucite has also been synthesized using a zeolite route, similar to the method used for creating the glass form of cesium. Irrespective of the method of forming pollucite powders, they could be pelletized by warm pressing prior to sealing in stainless steel capsules.

Production of cesium-137 pollucite at a facility that currently makes cesium chloride radiation sources would require establishment of a new process line, although the new line might be able to use some of the preexisting equipment. This would entail an investment that the committee is not equipped to estimate.

An alternative approach to reducing solubility and dispersibility is to make cement incorporating the cesium-137 by the addition of cement paste and fillers. This approach has the advantage of low-temperature processing and, with judicious choice of cement phase, low aqueous solubility (Mimura et al., 1997). However, the dilution associated with making cement limits the attainable specific activity. Also, the product remains a brittle solid that could degrade due to radiation effects, so it does not lower the cesium's potential dispersibility in an explosion. The cementitious approach has advantages for large-scale immobilization of wastes containing cesium-137.

It has been difficult to gather information on the details of production processes for cesium-137 sources at PA Mayak. A source distributor told the committee that there are difficulties in making high-activity vitrified sources, including self-shielding effects, although self-shielding is not a significant factor in these sources. Another expert alluded to radiation damage to the glass, which is more plausible (V. Pet'kov, Nizhni Novgorod State University, personal communication to M. D. Lowenthal, May 28, 2007). Researchers in Russia who work with PA Mayak informed the committee that although PA Mayak mainly produces cesium-137 sources in the chloride form, it also makes smaller quantities of both the glass and pollucite cesium sources. They are also working actively on developing phosphate ceramic forms (i.e.,

$CsMgPO_4$) that they say can potentially be made with specific and total activity more comparable to radioactive cesium chloride (B. F. Myasoedov, Russian Academy of Sciences, personal correspondence to M. D. Lowenthal, May 28, 2007; Pet'kov, personal communication, 2007). The committee's ability to learn about the Russian products, production methods, and research and development was somewhat limited—the committee is neither a potential customer nor a potential sponsor of work on these topics. An entity that can act in one of those roles could make more progress in gaining information and promoting the development of the more robust matrixes for cesium-137 (see Chapter 10 concerning research and development).

The price of cesium-137 chloride ranges from about $250/TBq ($9.30/Ci) for very large purchases (over 4,500 TBq [120,000 Ci]) to $10,500/TBq ($390/Ci) for a single relatively small (0.75-TBq [20-Ci]) source. A single 81.5-TBq (2,200-Ci) source costs a little over $34,000 (REVISS, 2003).

Many cesium-137 sources cannot be disposed in existing low-level radioactive waste disposal facilities. The Class C limit for cesium-137 is about 170 TBq/m^3 (4,600 Ci/m^3). The radiation sources and their packaging (over which volume the activity can be averaged) may be substantially smaller than a cubic meter, so the activity limit for disposal of a source is lower than 170 TBq. Further, just as with cobalt-60, waste-acceptance criteria restrict what sources may be disposed at a particular facility with much lower limits than concentration limits listed in the federal regulations. The U.S. NRC (2006a) reports that the Barnwell Disposal Facility in South Carolina does not accept cesium-137 sources containing more than 0.37 TBq (10 Ci). The U.S. Ecology low-level waste disposal facility in Richland, Washington, can dispose of a cesium-137 source of activity up to 1.1 TBq (30 Ci) in a 200-liter (55-gallon) waste drum, because that is the maximum allowed in the U.S. NRC Branch Technical Position on Concentration Averaging and Encapsulation (U.S. NRC, 1995)[10]

Some manufacturers that use cesium-137 will take back used cesium-137 sources for a fee. The cesium-137 may be repackaged for reuse or, if a Canadian company handles the decommissioning of the cesium source, the source may be disposed of in Canada. The OSRP had recovered 393 cesium-137 sources totaling about 310 TBq (8,393 Ci) as of January 2007, and another 600 sources totaling 905 TBq (24,446 Ci) were registered as excess or unwanted sources (Pearson, 2007).

Iridium-192

Iridium, one of the two densest metals (22.42 g/cm^3, same as osmium), is very hard and brittle, and is difficult to machine. It is also very resistant to chemical reaction and has a high melting point (over 2400°C). Natural iridium, which is found alloyed with platinum and in nickel ores, is 37 percent iridium-191 and 63 percent iridium-193. Iridium-192 radiation sources are used in gamma radiography (e.g., nondestructive inspection of pipes) and in brachytherapy, although brachytherapy sources in the United States are below the Category 2 threshold.

Iridium-192 radiation sources are made by irradiating natural iridium in a nuclear reactor. The iridium-191 can capture a neutron to create iridium-192, which has a 73.83-day half-life and has a 95 percent probability of decaying by beta decay to platinum-192 and emitting gamma rays and a 5 percent probability of decaying by electron capture to form osmium-192. In the

[10] According to the facility's 2007 rate sheet (US Ecology Washington, Inc., 2007), the marginal cost of disposal of a single drum could be as low as $20,630 if the radiation source is placed in a lead pig prior to solidification, reducing the dose rate at the surface of the drum below 10 mSv/hr (1 rem/hr). If a pig is not used, the cost is more like $64,310, based on a surface dose rate of 64 mSv/hr (6.4 rem/hr).

decay to platinum-192, on average 2.33 gamma rays are emitted with energies ranging from 135 keV to 1.378 MeV,[11] with an average energy of 380 keV.

Iridium sources are usually in the form of wires or stacks of thin foil discs rather than bulk material pellets, slugs, or powders. Used iridium-192 sources typically can be shipped back to their manufacturer and distributor or stored for decay because of their relatively short half-life. So, although disposal of iridium-192 is not a problem, the short half-life forces users to replace the sources frequently, meaning that many sources are in transport and storage at any given time.

Americium-241

Americium is an actinide or transuranium element with no stable isotopes. Like the other actinides, americium oxidizes fairly readily. Americium is produced by successive neutron captures in uranium-238, its activation products and decay products, to produce plutonium-241, which decays to americium-241 with a 14.4-year half-life. Americium is recovered from aging plutonium stocks in which it builds up through radioactive decay. Americium-241 decays with a half-life of 432.7 years by emitting an alpha particle. The alpha particle has an average energy of 5.465 MeV and is accompanied by a 13.9-keV x ray in 43 percent of decays and a 59.5-keV x ray in 36 percent of decays, and no x rays in the other decays. The decay product, neptunium-237, is also radioactive, with a 2-million-year half-life. Americium-241 is used both as an alpha source and with beryllium as a neutron source (called an americium-beryllium or Am-Be source). In an Am-Be source, some of the alpha particles from decay of the americium are absorbed in the beryllium, which then emits a neutron with energy ranging from 0 to about 11 MeV with the average energy at about 6 MeV. Am-Be produces about 1 neutron for 20,000 alpha decays.

Am-Be sources, such as those used in oil well logging, are typically formed by cold pressing mixtures of americium oxide (AmO_2) and beryllium powders to form a pellet which is then either diffusion bonded to a metal strip (for small sources) or sealed in a welded stainless steel container. Once supplied by the DOE Isotopes Program (separated from plutonium at Los Alamos National Laboratory), americium-241 is now only produced by the Russian radionuclide production facility at the PA Mayak. Americium-241 is supplied globally by PA Mayak, but through longstanding business agreements with PA Mayak and REVISS Services, the majority of Russia's americium-241 is made available to the West via QSA Global (formerly AEA Technologies QSA, Inc., formerly a part of Amersham). REVISS may also manufacture Am-Be well logging sealed sources directly.

The Am-Be material is sold by the gram (1 g is about 0.127 TBq [3.4 Ci]), but the price of a radiation source depends on several factors including the manufacturing batch size and particularly the customer's design specification regarding both physical/mechanical integrity requirements and neutron output. The prices are not public, but anecdotally the price range of a typical logging-while-drilling Am-Be well logging source now is approximately $30,000 to $50,000 and for a wireline Am-Be source it is $60,000 to $80,000. Because of the current shortage of Am-Be, there is a queue to get new sources as they come available. Some customers are willing to pay a premium above the current base prices for expedited delivery and other special services. QSA Global has said that the supply chain is responding to the Am-Be shortage by building a capability for more production. New Am-Be sources can also be obtained from old Am-Be sources. The "recommended working life" of an Am-Be source is 15 years, after which the source manufacturers recommend that the sources be recertified (if it is in good

---

[11] The 1.378-MeV gamma is emitted in one of every thousand decays. Two other, higher energy gamma rays (1.384 MeV and 1.406 MeV) are emitted in 3.2 and 3.9 of every 100,000 decays, respectively.

condition), reencapsulated (if the capsule is slightly damaged, but the design is still in use), or recycled (if the design is no longer in use or the damage to the capsule is severe, then the raw Am-Be can be removed and manufactured into a new source). Some companies, such as Gammatron, Inc., offer Am-Be source recycling services; QSA Global does not, although representatives of the company have suggested that it would if interest in the service grows.

Americium-241 sources cannot be disposed in commercial low-level waste disposal facilities. The Class C limit for alpha-emitting transuranic nuclides with half-life greater than 5 years is 3,700 Bq/g (100 nCi/g), a factor of nearly 300 million lower than the concentration in a typical radiation source. The only current disposition path for these sources is through the OSRP, which stores them pending approval to dispose of them. Sources that are determined to have originated in the DOE Isotopes Program (or its predecessors) can be disposed in the Waste Isolation Pilot Plant (WIPP), a deep geologic repository for transuranic waste of "defense origin." Other Category 2 americium sources have no disposal option available. As of January 2007, the OSRP had recovered or collected 10,154 of these sources, totaling 507 TBq (13,698 Ci). Another 1,012 sources totaling 62.6 TBq (1,689 Ci) have been registered with the OSRP as excess or unwanted sources (Pearson, 2007).

Plutonium-238

Plutonium is an actinide or transuranium element with no stable isotopes. It is a silvery-white reactive metal that turns a dull, darker hue when it oxidizes, which it does readily. It has low solubility in pure water, but saltwater and halide acids attack it vigorously. Plutonium-238 is produced by neutron absorption in neptunium-237, which itself is produced by irradiation of uranium in a reactor followed by chemical separations. Plutonium-238 has a half-life of 87.7 years, decaying by alpha decay with an average energy of 5.486 MeV. The decay product, uranium-234, is a naturally occurring radionuclide.[12] The heat generated by decay in relatively pure plutonium-238 is such that a solid sphere of the material the size of a golf ball will glow red from thermal radiation if it is not actively cooled.

The only current user of high-activity plutonium-238 sources in the United States is the National Aeronautics and Space Administration (NASA), which uses the sources in radioisotope thermoelectric generators (RTGs) for probes that require a significant nonsolar power source (e.g., the Cassini probe that was sent to the planet Saturn). The United States has some hundreds of kilograms of neptunium-237 stored in solution at the Savannah River Site. A facility for production of plutonium-238 and manufacture of the RTGs is being constructed at the Idaho National Laboratory. The heat sources are in the form of an oxide pellet with 1.25 TBq (33.6 Ci) of plutonium-238 constituting 80 percent of the metal in the oxide for a 1-W heat output (NASA, 2006). A source of this activity is estimated to cost $3,600 (NASA, 2004). The sources are loaded into a robust housing that is designed to keep them intact even in case of launch or reentry accidents.

Like americium-241 sources, plutonium-238 sources do not have a commercial disposal pathway. The OSRP has recovered 2,169 sources comprising 407 TBq (10,993 Ci) and has registered 112 more (298 TBq or 8,043 Ci) as excess or unwanted as of January 2007 (Pearson, 2007). The OSRP has disposed of many of the recovered sources in the WIPP.

---

[12] Uranium-234 constitutes 0.0055 percent of natural uranium because although it has a half-life of 245,000 years, it is also a decay product of uranium-238.

Selenium-75

Selenium is a volatile, reactive, and corrosive element chemically resembling sulfur and forming extremely toxic compounds. It has moderate density (4.3 g/cm^3 to 4.8 g/cm^3) and melts at 217°C. Selenium has several natural isotopes: selenium-74 (0.89 percent), selenium-76 (9.36 percent), selenium-77 (7.63 percent), selenium-78 (23.78 percent), selenium-80 (49.61 percent), and selenium-82 (8.73 percent). Selenium-75 decays by electron capture with a half-life of 119.8 days to stable arsenic-75, emitting an average of 1.75 gamma rays with an average energy of 215 keV each, and a peak energy of 800 keV. It is used in radiography cameras for thin-walled structures, although it is not commonly used in the United States.

Californium-252

Californium is an actinide element with no stable isotopes. It is produced by successive neutron captures in actinide targets. Californium-252 has a 2.645-year half-life and decays by spontaneous fission 3.1 percent of the time and by alpha decay in the other 96.9 percent. The fissions release neutrons, and thus californium-252 is a very intense neutron source (2.3×10^{12} neutrons per second per gram). Because a uranium-238 nucleus must absorb 14 neutrons without undergoing other reactions that reduce the number of nucleons to yield a californium-252 nucleus, californium is produced in very small quantities. Oak Ridge National Laboratory currently produces only about 0.25 grams of californium-252 per year from feedstock at the Savannah River Site. The Research Institute for Atomic Reactors in Dmitrovgrad, Russia, is the only other facility that produces this radionuclide, and its production capacity is estimated at 0.025 grams per year (NRC, 2003). Yet it takes only tiny quantities to make a useful source: A Category 2 californium source contains at least 0.2 TBq, which is only about 10 mg of californium-252. The OSRP has 16 californium-252 sources registered for recovery, totaling less than 37 GBq (1 Ci), and has already recovered 12, also totaling less than 37 GBq (Pearson, 2007).

Strontium-90

Strontium is a reactive metal typically found as an oxide or a salt. It has four stable isotopes, strontium-84, -86, -87, and -88, the last of which is the most naturally abundant (82.6 percent). The radionuclide strontium-90 is a fission product produced in 5.8 percent of thermal fissions in uranium-235 and 2 percent of thermal fissions in plutonium-239. Strontium-90 decays by beta decay (0.546 MeV) with a half-life of 28.78 years to yttrium-90, which itself decays by fairly high energy (2.28 MeV) beta decay with a 2.67-day half-life. Category 1 and 2 strontium-90 sources in the United States are only used as power sources in RTGs. Category 3 strontium-90 sources are used in radiotherapy of very superficial lesions.

Strontium 90 is generated in nuclear power or isotope production reactors and is found in high-level radioactive waste and fallout from nuclear weapons testing. Strontium-90 does not emit penetrating gamma rays, so when it is a contaminant it is only a concern for external exposure if it is deposited on the skin. The major concerns are internal exposures because of the high-energy beta emissions and because strontium is in the same chemical group as calcium, so the human body concentrates ingested strontium in the bones where it resides essentially permanently rather than being eliminated through common bodily functions. High-activity strontium-90 sources, however, produce significant bremsstrahlung radiation from stopping of the high-energy electrons emitted by nuclear decay. This bremsstrahlung radiation can be enough to cause deterministic health effects if a very high activity source is involved (such an incident occurred with an RTG in the Republic of Georgia in 2002).

The committee found no reliable cost estimates for strontium-90, although Malvadkar and Parsons (2002) report a cost found on the internet of $250 per watt thermal for a strontium-90 RTG. With about 5.4 TBq per watt, this translates to approximately $45 per TBq (about $1.70 per Ci) in an RTG that might contain over 1,500 TBq. As with cesium-137, strontium-90 can be acquired from the PA Mayak, in Russia. Also like cesium-137, there is a 259-TBq/m^3 (7,000-Ci/m^3) limit on the concentration of strontium-90 allowed in near-surface disposal of low-level radioactive waste. Thus, high-activity civilian strontium-90 sources have no disposal option available now. OSRP has recovered 10 strontium-90 sources containing nearly 2,740 TBq (74,000 Ci). Another 100 sources containing nearly 13,500 TBq (364,000 Ci) have been registered with OSRP for recovery (Pearson, 2007). Most of the activity in these registered sources is concentrated in a small number of RTGs.

USES OF CATEGORY 1 AND 2 RADIATION SOURCES IN THE UNITED STATES

The major applications for the high-activity sources include large panoramic irradiators for bulk sterilization, self-shielded irradiators for research and blood irradiation, teletherapy and Gamma Knife® machines for cancer therapy, RTGs, radiography cameras, and well logging tools. Each of these device applications is described briefly below. Some of the security hazards associated with the applications are discussed in the next chapter, and all but the RTGs are discussed in greater detail in Chapters 5 through 9.

Panoramic Irradiators

Among devices that use radiation sources, panoramic irradiators have the greatest activity. A typical panoramic irradiator contains 40,000 to 260,000 TBq (1 to 7 million Ci) of cobalt-60. These facilities are used primarily to sterilize single-use medical products and devices, but they are also used to sterilize other products. Panoramic irradiators are contained in large buildings with radiation shielding provided by a maze of concrete walls (the walls around the irradiation chamber are typically around 2 m [6 ft] thick). The cobalt-60 sterilization facilities use standard cobalt-60 pencils mounted in a source array. Each pencil contains 16 slugs of cobalt-60 with each slug approximately 8 mm in diameter and 2.5 cm in length (1/4 in. × 1 in.) having an activity of approximately 37 TBq (1,000 Ci). The pencils are clipped into a large, usually planar array that measures roughly 3 m × 6 m (10 ft × 20 ft) in size. The array is kept shielded in a pool that covers the array with several meters of water to serve as shielding when the array is not in use. Products to be sterilized are placed in containers that are carried by a conveyer (a belt or a hanger system) that passes the containers through the shielding maze and pauses next to the array for a set irradiation dose. In modern facilities, the entire process is automated. Panoramic gamma irradiators and their replacements are discussed in Chapter 6.

In 2002, the International Atomic Energy Agency (IAEA) reported that 142 sterilization and food irradiators operated worldwide (IAEA, 2002). This figure is probably now over 160 (see Chmielewski and Haji-Saeid, 2004), and there are 63 in the United States today. The cobalt-60 pencils come from two source suppliers: The Canadian company MDS Nordion has the largest share of the market, and the international marketing consortium REVISS has the rest. REVISS acquires its cobalt-60 from both PA Mayak and the Argentine CNEA (see Production of Category 1 and 2 Radiation Source Material, below).

Self-Contained Irradiators

Self-contained irradiators, also known as self-shielded irradiators, are used mainly for biomedical and radiation research, and blood irradiation. There are three self-contained irradiator device manufacturers operating today: MDS Nordion of Canada; CIS-US, Inc., a French company that no longer manufactures new machines but is still servicing those in existence here in the United States; and a U.S. company, J. L. Shepherd and Associates. There are 1,341 self-contained irradiators that use radionuclide radiation sources in the United States, approximately 85 percent of which use cesium-137, while nearly all of the remaining devices use cobalt-60 (U.S. NRC, 2007a).[13] These include blood irradiators, research irradiators, and calibration irradiators. Self-contained gamma irradiators and their replacements are discussed in Chapter 5.

The blood irradiators typically weigh about 1,000 kg and contain from 75 to 260 TBq (2,000 to 7,000 Ci) of cesium-137, with a typical source loading of 110 TBq (3,000 Ci). The U.S. NRC estimates that about 30 of the roughly 550 blood irradiators in the United States now use cobalt-60 (U.S. NRC, 2007a). (The use of cobalt-60 sources in self-contained irradiators is discussed in Chapter 5.) Self-contained irradiators used for research have a much broader spectrum of source sizes and device weights. The approximately 192 cobalt-60 research irradiator devices can weigh up to 4,000 kg and can contain over 1,100 TBq (30,000 Ci). The roughly 490 cesium-137 research irradiators can contain similar activity levels (740 TBq or 20,000 Ci) but weigh no more than 3,000 kg. This shows that cesium-137 dominates the research irradiator field, too, with over 72 percent of the roughly 680 machines using this radionuclide while nearly all of the rest use cobalt-60.

Some devices, including some kinds of radiation detectors and dosimeters, require irradiation calibration that is both precise and accurate at high doses. Some calibration sources used for these purposes are also considered self-contained irradiators. Radionuclide sources are typically used for this purpose because the decay energy (and decay rate) is known or readily calculable. These calibration irradiators are, on average, smaller-activity sources than the other self-contained irradiators, but some hold about 80 TBq (2,200 Ci) of cesium-137. The U.S. NRC reports 104 of these in non-fuel-cycle facilities, one of which uses cobalt-60 and all of the others of which are loaded with cesium-137.

Radiotherapy: Teletherapy and Gamma Knife®

Teletherapy devices are used to treat malignant tumors. Unlike brachytherapy, where small radiation sources are placed in or near the tumor, the teletherapy radiation source is kept at a distance from the patient and a beam of radiation is directed to the tumor. Teletherapy machines contain 37 to 550 TBq (1,000 to 15,000 Ci) and, at least in the United States, virtually all (247 of 248) use cobalt-60 (one uses cesium-137). Most of these are thought to be in storage for decay or converted to nonmedical uses (i.e., for fixed radiography, research irradiation, or teaching and research), because linear accelerators have replaced nearly all cobalt-60 teletherapy devices in medical practice in the United States, even in many veterinary clinics. There are an estimated 3,000 teletherapy machines in use around the world, and some of the older machines (e.g., the abandoned teletherapy device at Goiania Brazil) still contain cesium-137.

The Gamma Knife® competes with linear accelerator machines for the treatment of centimeter-sized brain tumors in areas of the brain where conventional surgery generally is not

---

[13] One device in the United States is classified as a self-contained irradiator with a californium-252 source.

possible. These machines are licensed to contain up to 245 TBq (6,600 Ci) of cobalt-60, encapsulated in 201 sealed sources each with an activity of up to 1.2 TBq (33 Ci). There are approximately 200 Gamma Knife® devices worldwide, including at least 104 in the United States. Elekta, a Swiss/Swedish company is the sole manufacturer of the Gamma Knife®, while MDS Nordion is the main source for the small cobalt-60 sealed sources. A Chinese company, GammaStar, has begun to market a competing device and Elekta is now selling a new version of the Gamma Knife® with 192 sources, instead of the 201 used in previous models.

RTGs

RTGs convert the heat produced from radioactive decay into electricity by the method of thermoelectric conversion. RTGs were used to provide electrical power to remote stations and for space missions. The older, legacy RTGs contain strontium-90, although NASA currently uses Pu-238 RTGs for special space missions. Both radionuclides generate relatively significant amounts of heat per decay. Large RTGs can contain several thousand TBq (hundreds of thousands of Ci), with the typical source size being 750 TBq (20,000 Ci) (IAEA, 2005b). The plutonium-238 RTGs made by DOE for NASA are produced, as needed, for specific NASA space missions. There are currently no commercial RTGs in the U.S. inventory, and the committee was told that the remaining U.S. RTGs are well secured. There are now international efforts to improve security and, in some cases, replace Russian RTGs with alternative technologies.

Well Logging Neutron Sources

The interaction of the neutrons from an Am-Be well logging source with the surrounding environment produces useful information about the geologic features through which the well was bored. The U.S. NRC's Interim Inventory reports 300 Category 2 well logging devices in the United States (U.S. NRC, 2007a). Many more are Category 3 devices. Americium-241 is now supplied exclusively by REVISS via the Russian radionuclide production facility at PA Mayak. QSA Global, Inc. is currently the sole manufacturer of Am-Be well logging sources. They receive partially fabricated Am-Be sources from REVISS and perform the final encapsulation into their sealed-source product. The three largest oil-field service companies, Schlumberger, Halliburton, and Baker Hughes, manufacture their own well logging tools in-house, incorporating the supplied Am-Be sealed sources. In addition to these "big three," there are medium-sized companies (e.g., Weatherford) and many smaller well logging oil-field service companies. Thermo-Electron is the largest manufacturer that sells well logging tools to these smaller service companies.

Industrial Radiography

Radiography devices are used to nondestructively examine the integrity of structures, manufactured components, metal forging, pipes, and welds, as well as fiber composites and composite structures. Many of the devices are portable and use radionuclides such as iridium-192, cobalt-60, selenium-75, ytterbium-169, and tellurium-170, with iridium-192 by far the most commonly used radionuclide.

The major manufacturers and distributors for industrial radiography devices include QSA Global, Industrial Nuclear, Source Production and Equipment Company, Agiris, and CIS-US. However, the actual producers of the radioactive material are MDS Nordion, PA Mayak, and the consortium of European reactors.

PRODUCTION OF CATEGORY 1 AND 2 RADIATION SOURCE MATERIAL

Figure 2-2 shows the method of production of the radionuclides in Category 1 and 2 sources in the United States, all of which are produced in nuclear reactors.[14] In reactors, the absorption of thermal neutrons by target material is used to produce radionuclides such as cobalt-60 and iridium-192. Fission products such as cesium-137 and strontium-90, and transuranics such as americium, plutonium, and californium, are generated by the capture of neutrons in a uranium target and are obtained by chemical processing of the irradiated fuel or target. The radionuclides of concern for use in radiological dispersal devices are generally produced in research reactors; however, a few commercial reactors are also used. There are over 250 research reactors currently operating worldwide, and 100 of these are involved in radionuclide production (DOE/U.S. NRC, 2003). However, only a few of these reactors are involved in major production for commercial/industrial/medical use. Note that radionuclide production in accelerators results in the production of small, short-lived radionuclides, none of which poses a significant terrorist threat.

COMMONLY USED RADIONUCLIDES IN THE U.S. INVENTORY

As noted at the beginning of Chapter 1, the best data on types and quantities of radiation sources in the United States available now are from the U.S. NRC Fiscal Year 2006 Interim Inventory of Radioactive Sources Data Analysis, which reports that there are 28,200 civilian Category 1 radiation sources and 25,532 Category 2 sources licensed by the U.S. NRC and Agreement States in the United States. These are estimates: The U.S. NRC staff estimates that there are 5,036 devices in total, and about 715 of these contain Category 1 sources and perhaps 4,320 devices contain Category 2 sources. The 215 devices that are listed having undetermined application in the summary of the 2006 Interim Inventory are counted here as containing Category 2 sources, but the committee was unable to check these numbers. These numbers do not include sources or devices at nuclear power plants or nuclear fuel cycle facilities, which were not included in the interim inventory (U.S. NRC is including them in the 2007 Interim Inventory). The estimated numbers of Category 1 and 2 devices in each state are presented in Figure 2-3. These data do not include manufacturers and distributors, fixed gauges, or devices for which the type was listed as undetermined.

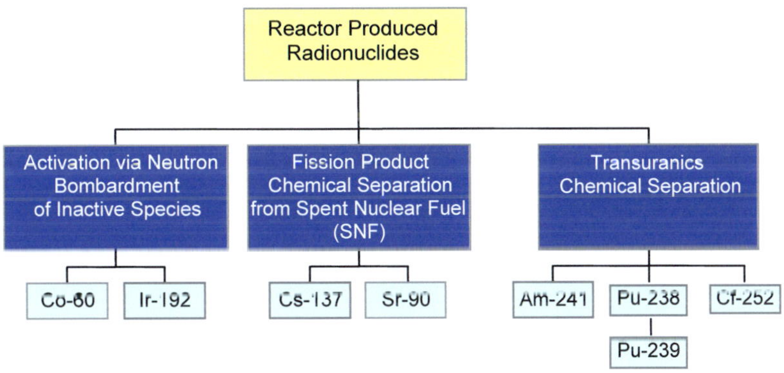

FIGURE 2-2 Origins of reactor-produced radionuclides for radiation sources. SOURCE: Adapted from Connell, et al. (2006).

---

[14] A good reference on the topic of radionuclide radiation source production and use is by Ferguson et al. (2003).

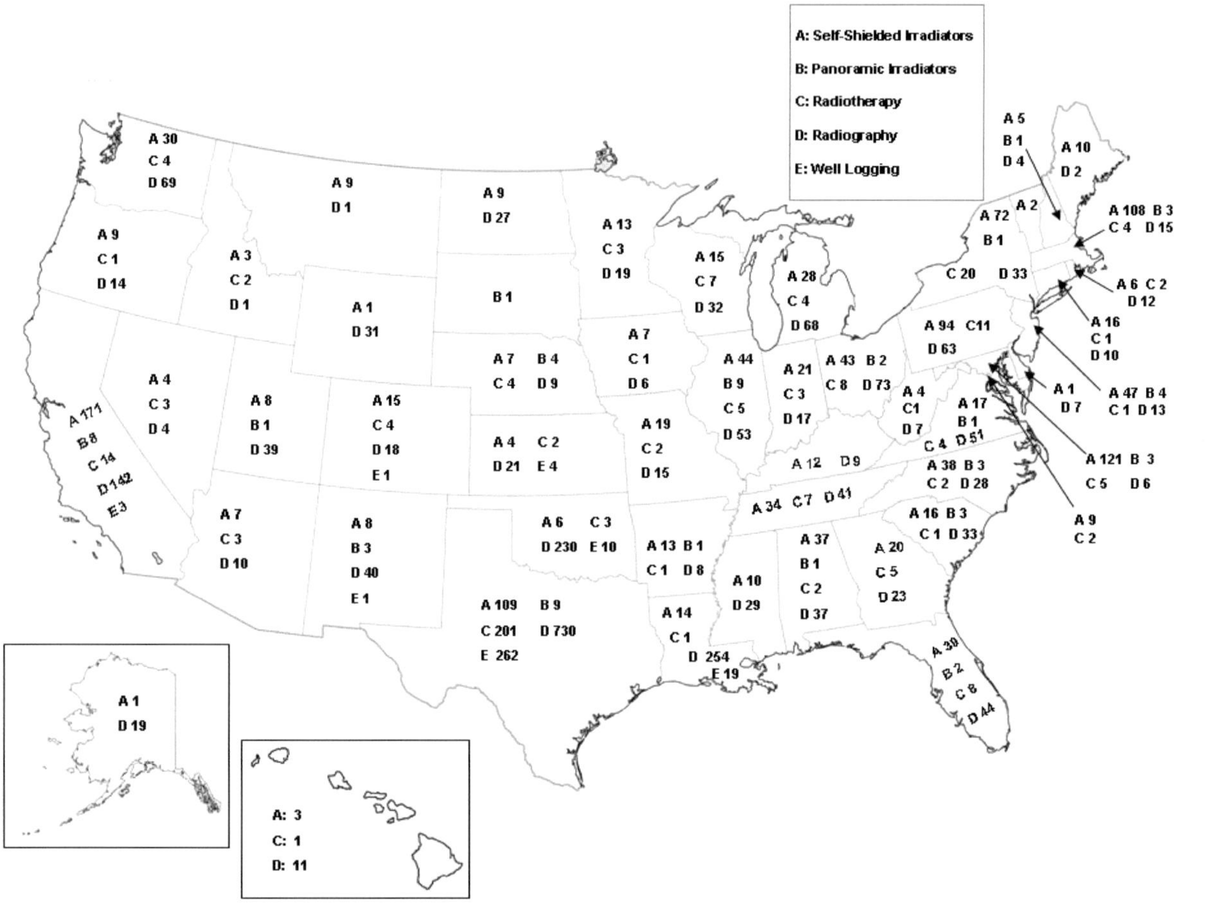

FIGURE 2-3 Estimated numbers of Category 1 and 2 devices broken down by state. These data do not include manufacturers and distributors, fixed gauges, or devices for which the type was listed as undetermined. Locations are mostly based on the contact location listed on the license. SOURCE: Constructed with data from U.S. NRC (2007a).

The method used to collect the data—a voluntary survey of U.S. NRC and Agreement State[15] licensees reporting only what Category 1 and 2 sources were in inventory at the date of their response to the survey—does not ensure that every source is accounted for and does not track where the sources are. When the U.S. NRC imposed increased controls on Category 1 and 2 licensees, the agency examined every materials license, which resulted in identification of 545 additional U.S. NRC licensees who are licensed to hold Category 1 or 2 radiation sources but had not been surveyed in previous years. Only 10 percent of those actually had Category 1 or 2 sources in their possession when surveyed.

The Agreement States verified to the U.S. NRC that they checked their own licensee lists. U.S. NRC staff members expressed their concern that, in the absence of a source tracking

[15] Section 274 of the Atomic Energy Act of 1954 as amended authorizes the U.S. NRC to "enter into agreements with the Governor of any State providing for discontinuance of the regulatory authority of the Commission … with respect to [explicitly identified] materials within the State …. During the duration of such an agreement it is recognized that the State shall have authority to regulate the materials covered by the agreement for the protection of the public health and safety from radiation hazards." States that have entered into such an agreement with the U.S. NRC are called Agreement States. The U.S. NRC retains authority in these states over nuclear fuel cycle facilities, export and import of nuclear materials, and radioactive waste disposal (as determined by the U.S. NRC).

system, some people are trying to extract more information from the Interim Inventory than the surveys were meant to provide. Fuller and more accurate data are indeed expected from the National Source Tracking System, which is scheduled to come online in the summer of 2008, but the data available now are adequate for the committee's purposes.

To appreciate the prevalence of different radionuclides in commercial use, one can look at the total activity in Category 1 and 2 radiation sources by radionuclide (see Figure 2-4) and the numbers of Category 1 and 2 devices by radionuclide (see Figure 2-5 and Table 2-1). There are over twice as many Category 2 devices as there are Category 1 devices, but the Category 1 devices comprise 99 percent of the total activity. Nearly all of the activity (96 percent of the total for all sources) is in the form of cobalt-60 sources at 62 panoramic irradiators used primarily for sterilization (see Chapter 5). Most of the rest is in self-contained irradiators (1,117 that use cesium-137 and 224 that use cobalt-60; see Chapter 4) and radiotherapy devices (351 that use cobalt-60[16] and 1 that uses cesium-137; see Chapter 6). There is only one panoramic irradiator that uses cesium-137, and it is a dry storage panoramic irradiator. U.S. NRC actions on the use of radioactive cesium chloride sources in panoramic irradiators, even dry source irradiators, indicate the commission's skepticism about the use of such sources in panoramic irradiators. (See the discussions of cesium chloride in panoramic irradiators in Chapters 3 and 10.)

The committee has not examined RTGs, and so has excluded the 24 Category-1 and 10 Category-2 strontium-90 devices and any plutonium-238 devices from discussions in later chapters. RTGs are a concern in the republics of the former Soviet Union, where over a thousand of the Category 1 devices were produced, many of which are stored with little or no security in place. The Category 1 RTGs in inventory in the United States are not a major concern in the committee's view because there are few of them and they are stored in secure government facilities only for military applications. The U.S. government also produces plutonium-238 RTGs for deep-space missions (probes to the outer planets). None of these is listed in the inventory (most are Category 1 devices and can contain several thousand TBq), but they are produced by DOE and delivered to NASA just prior to launch of the probes.

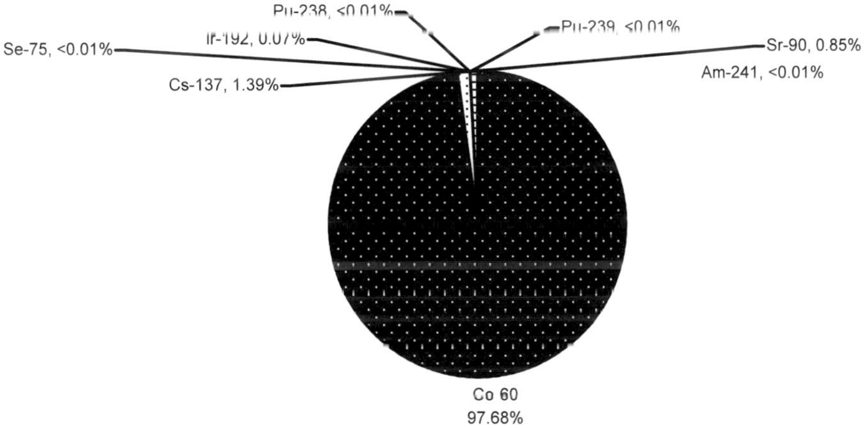

FIGURE 2-4 Share of total activity in radiation sources in the United States by radionuclide. SOURCE: Constructed with data from U.S. NRC (2007a).

---

[16] This number includes 104 Gamma Knife® devices and 247 radiotherapy devices. The latter number reported in the U.S. NRC 2006 Interim Inventory (U.S. NRC, 2007c) is much higher than the number of devices currently used for teletherapy in the United States, perhaps by a factor of 10. Many of these are in the process of storage for decay at a centralized facility, and others are being used for other irradiation applications.

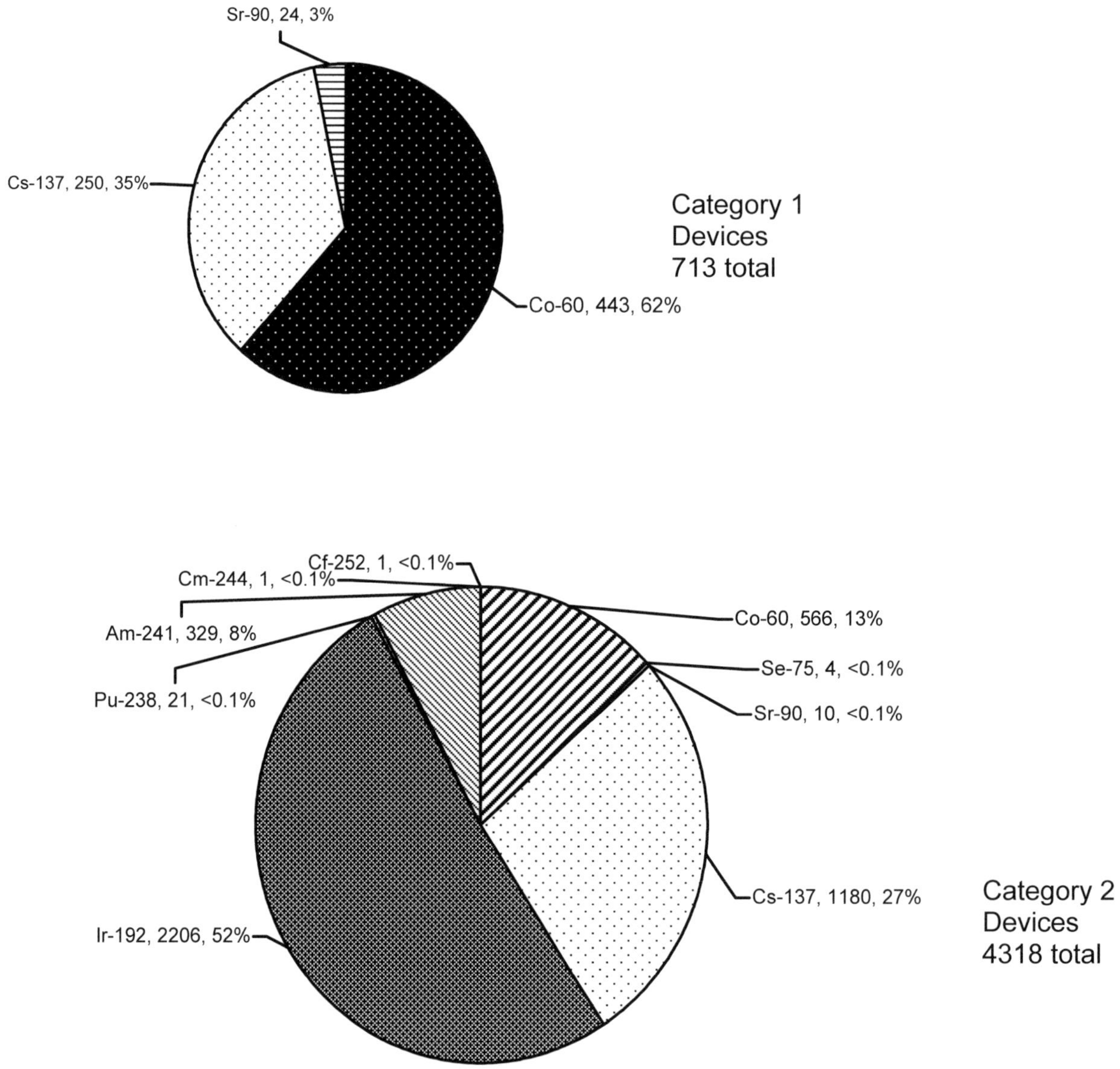

FIGURE 2-5 Numbers of Category 1 and 2 devices in the United States. Each identifier lists the radionuclide, the number of devices, and the percentage of the total number of devices in that category. The areas on these pie charts are approximately normalized to the same scale, so that equal areas represent equal numbers. NOTE: Several assumptions have been made about the IAEA categories of devices that are not explicitly categorized in the Interim Inventory Summary. SOURCE: Constructed with data from U.S. NRC (2007a).

RADIATION SOURCE PRODUCTION AND DISTRIBUTION

One can visualize the international supply chain for radionuclides as a pyramid system. At the top are a small number of organizations that have access to nuclear reactors and produce radionuclides in significant quantities.

A summary of the main producers of radioactive material in the world is shown in Figure 2-6. The two largest producers of the commercial radionuclides are MDS Nordion of Canada (with the largest share of the cobalt-60 market) and REVISS, an Anglo-Russian consortium. Argentina, through its National Atomic Energy Commission (CNEA) also produces some cobalt-60 which it is on contract to sell to REVISS. MDS Nordion contracts with the Canadian government's Chalk River Nuclear Laboratory and with nuclear power utilities to have radionuclides produced in their reactors. REVISS acquires all of its supplies of some radionuclides from the PA Mayak. Other radionuclides are acquired from several sources.

The old U.S. Atomic Energy Commission (AEC) was once very active in producing and distributing radionuclides for commercial, medical, and research applications initiated as part of the Atoms for Peace Program in the 1960s. The U.S. government, through the DOE Isotope Program (which inherited the older AEC program) ceased supplying cesium-137 many years ago and just recently ceased supplying americium-241, leaving REVISS as the sole supplier of these two radionuclides. The DOE Isotope Program now mainly produces relatively short-lived radionuclides for research and medical applications, although it does also produce californium-252. International Isotopes, Inc., contracts with DOE to produce cobalt-60 sources.

A small number of European research reactors are used to produce iridium-192. There are other, regional, suppliers of radionuclides, such as the Eastern European, South African, Indian, and Chinese producers (Van Tuyle et al., 2003). To the committee's knowledge, they do not supply outside their regions and do not yet serve in the United States, although they might in the future.

At the next level of the pyramid are the manufacturers and distributors that manufacture sealed sources from the radionuclide product (unsealed radiation sources) or repackage encapsulated sources and place them into devices for purchase by licensed users. There are approximately two dozen manufacturers and distributors in the United States that receive or hold Category 1 and 2 source materials. In some cases the radionuclide producer is vertically integrated with the device manufacturer. That is, the same company produces sources and manufactures devices, as is the case with MDS Nordion.

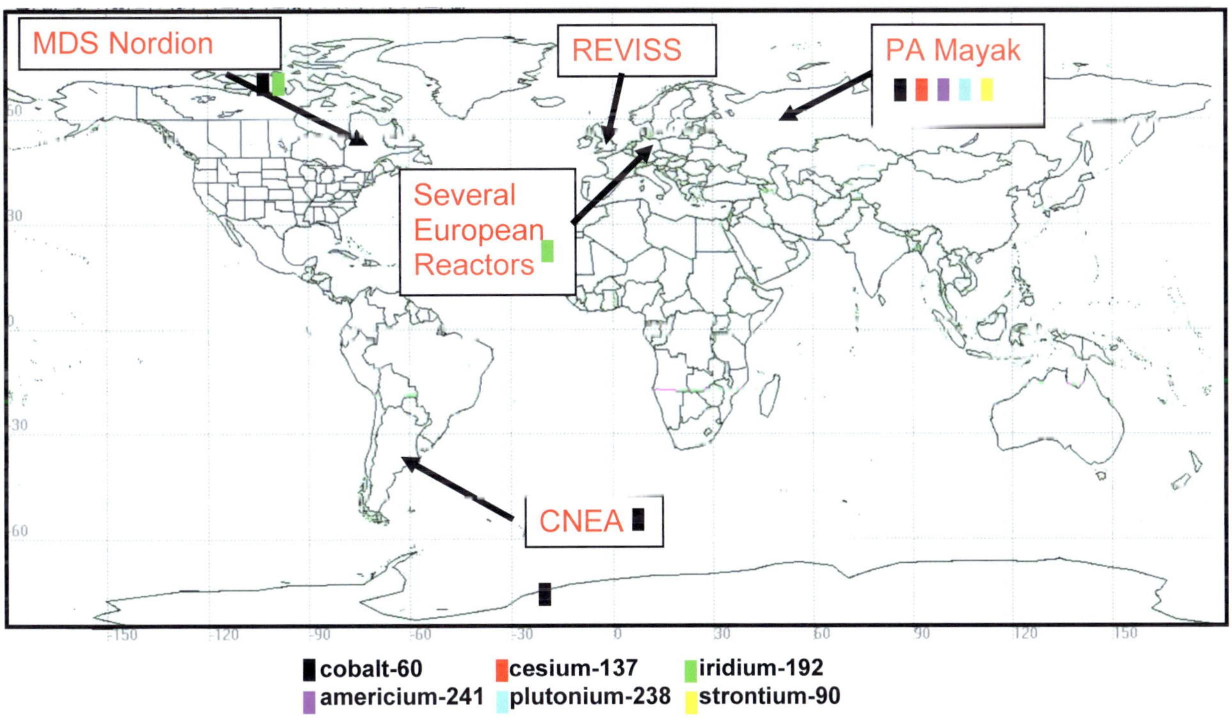

FIGURE 2-6 Major international radionuclide producers. SOURCE: Adapted from Van Tuyle, G., et al. (2003).

At the bottom of this supply pyramid are the licensed users. According to the IAEA, there are more than 20,000 users of what the IAEA considers high-risk radioactive sources and devices (IAEA, 2002) across the world.

An examination of the U.S. inventory of radiation sources shows that the top four radionuclides (in decreasing order based on total activity) are cobalt-60, cesium-137, americium-241, and iridium-192. However, simply examining the inventory quantity alone is not sufficient when assessing the risk posed by these radionuclides if used for malevolent purposes or for evaluating options for implementing replacements of high-risk sources. These risk considerations are discussed in Chapter 3.

FINDING AND RECOMMENDATION

Finding: The radionuclide radiation sources examined in this study are used in applications that are important to the nation's health, safety, and economic strength.

High-activity radiation sources are used in the United States and other modern societies in a variety of ways: they are used in devices that improve the success of medical procedures—ensuring that medical devices and implants are sterile, preventing fatal complications from bone marrow transplants, and providing noninvasive techniques for treating brain lesions; they are used in devices for inspecting the integrity of buildings, bridges, and industrial equipment; and they are used to seek out oil and gas resources deep in the ground. These applications are immensely valuable to the United States. The question is not whether these activities should continue, but whether lower risk replacements for the radiation sources are feasible and practical, and what steps should be taken to implement replacements for the sources that pose a high risk to public health and safety.

Recommendation: Replacement of some radionuclide radiation sources with alternatives should be implemented with caution, ensuring that the essential functions that the radionuclide radiation sources perform are preserved.

As the nation seeks to improve safety and security, the value and benefits of current practices should be recognized and, where possible, the services the devices provide should not be compromised. Some replacements do entail trade-offs with respect to safety, security, costs, convenience, and performance, as discussed in Chapters 3 through 9. These trade-offs should be considered carefully. A reduction in the performance of a device may be acceptable if it provides sufficient benefits in safety, for example. Replacement should preserve acceptable performance of these applications to preserve the benefits that these applications provide, on many of which the United States has come to rely.

CHAPTER 3

RADIATION SOURCE RISKS

SUMMARY

This chapter presents a discussion of hazards and, to some extent, risks associated with uses and misuses of the radiation sources described in Chapter 2. The uses are described in detail in Chapters 5 through 9 with examinations of possible replacements.

The committee used a qualitative risk framework in thinking about risks. The use of one or more radiation sources as radiological dispersal devices (RDDs or "dirty bombs") to cause lasting contamination that prevents regular human access to an area (area denial), and the economic and social consequences that result may be the most important risks of malevolent use of radiation sources. The International Atomic Energy Agency (IAEA) categorization of radioactive sources and the corresponding U.S. Nuclear Regulatory Commission (U.S. NRC) categorization are based almost exclusively on deterministic health effects from exposure to an unshielded source, not on the area-denial RDD potential of the radionuclides. Although these categorization systems provide a basis for regulatory regimes related to safety and some aspects of security, they do not account for other important security-related issues. In this chapter, the committee recommends that the U.S. NRC reexamine its security-based orders and decisions for materials licensees, considering the potential consequences of area-denial RDDs.

For the purpose of prioritizing radiation source replacements and considering options for implementing those replacements, the committee concludes that radioactive cesium chloride is the greatest concern among the materials used in radiation sources in the United States because it is used in significant quantities in urban areas in a powdered, dispersible form. Review of previous accidents involving cesium-137 dispersals shows significant consequences. After an incident with radioactive cesium chloride in a wet panoramic irradiator, the U.S. NRC imposed a "qualified ban" on radioactive cesium chloride sources in such applications. In this chapter, the committee finds that radioactive cesium chloride sources should be replaced and that government action is needed to implement such replacements. The committee suggests several steps to implement the options for replacement.

RISKS AND PRIORITIES

The committee's charge (Sidebar 1-1) explicitly directs the committee to focus on risk: "The report will contain a review of radiation source use, potential replacements for sources that pose a high risk to public health or safety, and findings and recommendations on options for implementing the identified replacements." The U.S. NRC helped the committee to interpret the charge by stating at the committee's first meeting that "options for implementing" include both technical and policy options. The U.S. NRC wrote into its request for this study the description of Category 1 and 2 sources as radiation sources that pose a high risk to public health or safety.[1]

---

[1] Sources that fall into Category 3 and lower can be assembled into Category 2 or 1 quantities of radioactive material. Further, it may be the case that some radiation sources near the upper threshold for Category 3 pose more serious risks than other sources that fall near the lower threshold of Category 2 in

The committee chose to base its findings and recommendations on the hazards of the radionuclide radiation sources and the risks they pose, and to use these factors in prioritizing and balancing among trade-offs involved in radiation source replacement.

The term risk is used imprecisely and often inconsistently in common parlance and sometimes even in technical discussions. The committee uses the risk terminology as defined in a 1983 National Research Council (1983) report and other basic references (e.g., Kaplan and Garrick, 1981). In this framework, a hazard is a potential source of a negative consequence or harm, and risk is defined as the likelihood that such a harm will occur. The risk associated with a particular event involving a hazard (e.g., someone encountering the hazard posed by a knife) is expressed as the product of the probability of the event (someone encountering the hazard by being cut by the knife) and the consequence (a laceration):

$$\text{Risk} = \text{Probability} \times \text{Consequence} \qquad (3\text{-}1).$$

No single result characterizes the risks associated with a radiation source because there are many possible events involving a radiation source that have consequences, so it is more informative to consider an ordered set of risks. The events may be accidents or malevolent uses of the radiation source for which data may be poor and rigorous quantification may be difficult or impossible, particularly in characterizing the probability.

In the context of terrorism and other malevolent misuses, it may still be possible to evaluate consequences (in terms of the number of fatalities, economic losses, and social effects) for specific scenarios with relative rigor. Evaluation of probabilities, however, lies beyond the ready reach of traditional analytic techniques because the probability of a successful terrorist attack involves many factors that cannot be objectively quantified. An assessment that examines only the consequences is called a “hazard analysis” or “hazard assessment.” A hazard assessment can be informative, but is not usually used as the sole basis for risk management, because it can lead to inefficient and inappropriate allocations of resources that can increase rather than reduce risk. Stated differently, risks can actually increase if all of society’s risk mitigation efforts are devoted to events that have high consequences but very low probabilities, neglecting lower consequence, but higher probability events.

A formalism is emerging for evaluating risks related to terrorism, using heuristics (rules or devices used to narrow the scope of intractable problems) and whatever other data are available for systematic treatment of the problem (see, e.g., Haimes, 2006; Willis et al., 2005; Paté-Cornell and Guikema, 2002). In this approach, the probability term in the risk equation is usually described as the product of two probabilities: one characterizing the threat and the other characterizing the vulnerability. The threat to a target is a measure of the existing intent and capability to cause harm or damage by carrying out an attack, expressed as the probability of a particular kind of attack on a specific target in a given time period. A target’s vulnerability to the threat is a set of conditions or states of a system that can be exploited to harm that system. It is expressed as the probability that damage occurs from a given threat. The probability in the risk equation would then be the product of these two probabilities.

The committee was equipped to evaluate hazards or, from the risk formalism, some of the potential consequences, parts of the vulnerabilities, and none of the threats associated with misuse of radiation sources. Further, it is the role of policy makers to decide what levels of risk are acceptable and how to achieve those levels (e.g., through deterrence, detection, denying access, interdiction, mitigation, or reduction of the hazard). At the same time, however, the committee and the readers of this report can use the concept of risk qualitatively and heuristically to help organize thinking and better target society’s response to the threats. Unless

---

scenarios other than those used to create the source categorization system. However, the examination of Categories 3 to 5 is beyond the scope of this report.

otherwise indicated, wherever the term risk is used hereafter in this report, it is meant to connote the portions of risk (consequences and some aspects of vulnerability) that the committee could examine.

The risks associated with radiation sources include both the risks from accidents (i.e., equipment failures and unintentional misuse) and the risks from malevolent uses (i.e., sabotage or weaponization). The former can be called safety risks and the latter called security risks. Each of these is discussed below.

Radiation Source Safety Risks

Safety has historically been the U.S. NRC's main focus in regulating the possession and use of radiation sources. Most of the agency's and the Agreement States' regulations, guidance, and enforcement center on ensuring radiation safety, which they generally do quite well. Examination of the summaries of radiation incidents in the United States[2] shows that accidental exposures do occur, most commonly when required radiation safety procedures are not followed, and the doses received are typically well below 10 mSv. Most of these are direct exposures to gamma radiation when a source is not where it should be (such as a radiography source that does not retract into its housing) or when a person is in an area where he or she should not be (someone wanders into close vicinity of an active radiography camera missing warnings or because warnings are not in place). Major accidental radiation exposures—ones that could cause serious injury or death—are rare.

Radiation Source Security Risks

Security, too, has always been part of the U.S. NRC's mission, and many of the commission's regulations serve both safety and security goals. Accidental exposures resulting from inadvertent access have resulted in some of the worst incidents involving radionuclide radiation sources, so many of the commission's security measures were designed to prevent inadvertent access. Since September 2001, in response to the changing threat environment, the U.S. NRC's regulatory focus has sharpened to give security risks related to malicious acts a much greater emphasis than they had previously. The agency has taken steps to prevent and mitigate many security risks, including radiological terrorism. There are numerous scenarios for radiological terrorism, but all can be grouped into three categories (see, e.g., Moore, 2003):

- exposing people to radiation by using a radiation exposure device (RED) involves the clandestine placement of a large radiation source in an area where large numbers of people are likely to be exposed;
- poisoning food or water supplies with radioactive material; or
- dispersing radioactive material either through sabotage of a device in place or by fashioning and operating an RDD.

Each of these kinds of attacks can result in fatalities. The committee notes, however, that it is not easy to cause a large number of deaths with radiation sources, regardless of the kind of attack, and there are more direct pathways to lethality (such as bullets and bombs) than using radiation sources. However, our society also values more than just health and safety, and so those who seek to harm us can attack in other ways, aiming to harm our sense of well-being

---

[2] These are available from the state regulatory agencies and the U.S. NRC. See, e.g., http://www.dshs.state.tx.us/radiation/pdffiles/is1q03.pdf (accessed May 24, 2007).

and our economic prosperity. Psychosocial effects and economic damage, then, are important consequences to consider.

REDs

Time, distance, and shielding are the key elements of radiation protection and minimizing exposure. REDs use the same elements with an opposing goal: They cause more harm if people are exposed for longer times, at closer proximity, and with less shielding material between the source and the subject. These factors make it difficult to do grave harm to large numbers of people because it is difficult to put many people in close proximity to a source for a long time, and even the human body itself provides some shielding, so that a crowd somewhat shields a radiation source. A single RED, even a large one, might not have major or lasting psychosocial or economic impacts. Multiple REDs that target some critical element of society could have greater consequences than a single attack, but mitigation strategies are readily available, because REDs are inherently easy to detect with radiation detectors.

There are a large number of portable radiography cameras at job sites and in transit between licensees. These sources appear to be the most common Category 1 or 2 sources involved in accidental exposures, and RED attacks closely resemble the most common accidental exposures because they involve a person who is unaware that he is in the vicinity of a gamma radiation source.

Radiation Sources as Poison

Poisoning with radioactive material has garnered some attention since the murder of a former Russian KGB agent in London with polonium-210 (Po-210) in 2006 (see, e.g., Roessler, 2007), but it is difficult to conceive of high-impact physical or economic consequences using this kind of attack. Poisoning large numbers of people to achieve near-term health impacts would be difficult because in food, for example, the radioactive material must be fairly highly concentrated to have a deterministic effect on any individual—bacteria are much more effective at causing harm—so to affect many people requires a very large amount of material. Soluble radioactive material could be introduced into water reservoirs, but almost any plausible number of radiation sources would become too dilute to have much health impact. The material could be introduced closer to the point of consumption, but then the number of people affected would be low. It is possible that a poisoning attack could trigger some mistrust of the food or water supply, but because food-borne and water-borne illness outbreaks occur with some frequency, they are somewhat familiar. Problems with spinach and pet food in 2006 and 2007 have caused temporary economic damage and some concern about food safety, but there is no indication that suppliers of these products will suffer enduring harm. It is also possible that if a reservoir were contaminated with radioactive material, consumers would insist on cleaning up the reservoir, even if the radioactive material had no safety implications for the water in people's homes. Such cleanup could be costly, but such an attack would no longer be about poisoning; it would be a use-denial radiological dispersal attack.

RDDs

As with the other modes of attack, it is very difficult to cause serious deterministic health effects for large numbers of people with an RDD, even a very large RDD (Harper and Musolino, 2006). Just as with an RED, time, distance, and shielding are important, and people can evacuate or shelter in shielded areas, but the concentration or intensity of the source is another important factor. A Category 1 or 2 source is required for an RED to cause any serious deterministic health effects from external exposures. Dispersing radioactive material reduces its concentration, which lowers the likelihood of deterministic health effects from external exposure. People can also get harmful internal exposures from inhaling radioactive particles but, as with poisoning, deterministic effects require fairly high concentrations of the radioactive material, in this case in the form of respirable particles. Dispersing radioactive material increases the likelihood that people will be able to inhale the material, but again, as its concentration goes down, so does the likelihood of deterministic health effects.

Dispersal of radioactive material can create persistent area contamination. If dispersed at the right concentrations, the contamination may prevent people from occupying or even using the affected area. An RDD that maximizes this consequence is termed an "area-denial" RDD. RDD risks are decomposed in Figure 3-1. The figure shows a simple block diagram of the major components of RDD risk, dividing it into the two major elements: probability of the event and consequences given a "successful" attack.

For an RDD attack to occur, there must be someone or some group sufficiently motivated to undertake it. That group must obtain a quantity of radioactive material, build an RDD, and successfully deliver it to a target. These are the fundamental building blocks or elements that the terrorists require to carry out an RDD attack. The consequences of a successful RDD attack are listed beneath the Consequences box in the figure. They include health effects to the people exposed, economic damage from the area contaminated, and the less tangible psychological and social impacts. These consequences are not totally independent: prompt health effects and the potential for such effects lead to both psychological and economic consequences; specifically, fears and anxieties can cause large adverse economic consequences. However, the quantity of radioactive material needed to cause significant consequences varies according to the type of consequence. As noted earlier, it is very difficult to cause serious deterministic health effects for large numbers of people with even a large RDD. At the other end of the spectrum, even small or ineffective radiation dispersals may stimulate a psychological impact or a government response (*Nucleonics Week*, 1999).

Past accidents have demonstrated that radiation, due to its nature and history, has a unique ability to trigger fear and anxiety in the general population (IAEA, 2006, 1991, 1988). The quantity of radioactive material required to cause economic consequences is somewhere between the large amount needed to cause deterministic health effects and the small amount that might trigger psychological impacts. What is needed to cause economic impacts and the scale of those impacts are discussed below.

The federal government has developed plans to deal with natural disasters, accidents, and attacks covering a broad set of incidents, including nuclear and radiological incidents. An annex to this National Response Plan (DHS, 2004) establishes roles and responsibilities for nuclear and radiological incidents. Supporting these plans are proposed guidelines for the levels of contamination that would cause the U.S. government to relocate the inhabitants of a contaminated area and to initiate a cleanup campaign (*Federal Register*, 2006). The guideline for relocating inhabitants in a contaminated area is 20 mSv (2 rem) in the first year after the incident. That is, if estimates show that an inhabitant who continues to reside in the area would receive 20 mSv in the first year, the inhabitant should be relocated.

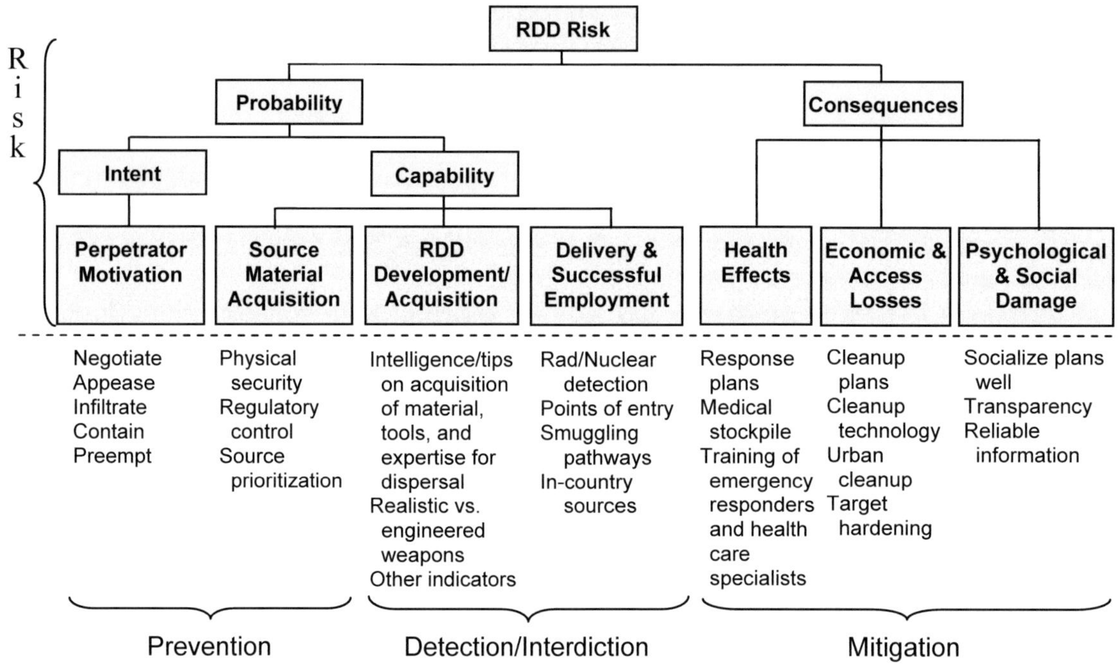

FIGURE 3-1 RDD Risk Decomposition. SOURCE: Adapted from Connell et al. (2003).

Twenty mSv is much lower than the doses required to cause deterministic health effects (immediate or near-term effects, such as those considered in establishing the IAEA source categories). Current best estimates indicate that 20 mSv corresponds to an increase in the average lifetime risk of cancer for an adult from about 42 percent (the current average risk of cancer) to about 42.2 percent (National Research Council, 2006b).[3] Under the guidelines mentioned above, the long-term cleanup level or extent of decontamination is to be determined through a case-specific "optimization process." If history and other contamination incidents are any guide, the cleanup level will be pushed much lower than this: most action levels for cleanup of contaminated sites under Superfund regulations are set at or below the level of 10^{-4} (1 in 10,000) lifetime risk of excess fatal cancer, which is well below the level of 20 mSv/year based on the linear, no-threshold model of radiation effects.

Referring back to Chapter 1, the economic consequences of an area-denial RDD attack are captured by the TRC term:

$$TRC = P\ (RDD) \times \text{Economic Cost (RDD)} \qquad (3\text{-}2).$$

Equation 3-2 states that the terrorist contribution to the cost of using a radiation source is the probability that a terrorist will successfully cause an RDD incident involving this source (P(RDD)) multiplied by the economic consequence of the dispersal of radioactive material (Economic Cost (RDD)). Thus, to actually derive a number for TRC, we need not just the economic cost of the dispersal but also the likelihood that it will occur. Unfortunately, both terms are difficult to estimate. Indeed, we have already discussed that the P(RDD) term is not readily

[3] The BEIR-VII lifetime risk model predicts that approximately 1 person in 1,000 would develop cancer from a 10 mSv one-time dose. Some 420 of those same 1,000 people would be expected to develop cancer from other causes (National Research Council, 2006c). Note that mortality from these cancers is perhaps one-half these numbers, depending on age at exposure and several other factors.

quantified and is itself the product of the probability of an attack and the probability that the attack will result in a particular consequence.

Only a few reports have been written on the economic costs of radiation dispersals. Those prior to the September 11, 2001, attacks on the United States (9-11 attacks) focus mainly on plutonium dispersal incidents when nuclear weapons were accidentally dropped from aircraft but did not detonate with a nuclear yield (see, e.g., Chanin and Murfin, 1996). Post 9-11 reports are more directed at RDDs (see, e.g., Reichmuth et al., 2005; Kelly, 2002). More economic studies are likely in development, in view of the current interest in and uncertainties about RDD consequences. The existing reports indicate that the cost of cleanup is an exponentially increasing function of the decontamination factor (DF)[4] needed to reduce the existing contamination level down to whatever the public and government mutually agree is clean enough for re-inhabitation of the area. The DF required for a cleanup depends on the level of contamination caused by the radiation dispersal and the cleanup standard. A report by the National Radiation Protection Board in the United Kingdom (NRPB, 1996) states that common "muck and truck" methods of cleanup (i.e., large-scale removal of contaminated material, such as such as sweeping, vacuuming, hosing with water, brushing, and application and removal of strippable coatings) are able to achieve DFs on the order of 10 (i.e., they are able to reduce the contamination level by a factor of 10) although factors of 3 to 5 are more common. The same source states that sandblasting can result in a DF of up to 100 for smooth surfaces, if it is done within 30 days after the dispersal. Over time, the dispersed material diffuses into the surfaces on which they were deposited, making the removal more difficult and reducing the effectiveness of sandblasting dramatically (a factor of 10 or more). According to Reichmuth et al. (2005), achieving a DF of 100 or more is not generally possible except by destructive methods (demolition of the contaminated structures and removal of the debris).

The total cost of cleanup includes both direct and indirect costs. Direct costs include the physical cost of the cleanup operation, the cost of disposal of the radioactive debris, and the cost of compensating individuals and business that were forced to relocate outside the contamination zone. Reichmuth et al. (2005) estimate that, for a highly contaminated zone requiring a DF greater than 10 in a high-density urban area, the direct costs would range from \$10 billion to \$40 billion/km^2. This could be consistent with figures in the NRPB report (1996), depending on the extent of contamination and the nature of the environment contaminated.

Indirect costs include the overall impact on the nation's economy resulting from the lost business in the affected zone and how these losses ripple through the economy, causing other losses. Rosoff and von Winterfeldt (2007) carried out an analysis of economic consequences of RDDs on the ports of Los Angeles and Long Beach. Their estimates of cleanup costs range from hundreds of millions to tens of billions of dollars, but these were based on the same work of Reichmuth et al. (2005). Their estimates of the indirect costs resulting from port shutdown and related business losses similarly range from hundreds of millions to tens of billions of dollars, depending on the magnitude of the attack. The committee has not reviewed the studies of direct or indirect costs in detail and so draws no conclusions about the reliability of the results, but notes that these are commonly cited figures.

To appreciate the hazards associated with Category 1 and 2 radiation sources, one has to understand the actual devices in which the sources are used. Each of the major applications of Category 1 and 2 radiation sources is described briefly below. Chapters 5 through 9 describe these applications in more detail from the perspective of function and possible replacements.

[4] The ratio of the before-and-after contamination levels establishes the decontamination factor.

Panoramic Irradiators

Panoramic irradiators are somewhat self-protecting against attacks that require human proximity because exposure to a 37,000-TBq (1-million-Ci) cobalt-60 source (at 1-m separation) would result in an incapacitating dose in about 10 seconds. Furthermore, the thick concrete structure provides additional security from sabotage attacks, and there are Compensatory Measures (special security requirements) mandated by the U.S. NRC at all of the large U.S. panoramic irradiator sites. The sterilization irradiators do, however, require re-sourcing at least once per year, which involves the transport of large quantities of cobalt-60 throughout the United States and installation of cobalt-60 pencils in the source rack. Requirements for radioactive material quantities of concern (RAMQC) apply to these shipments.[5]

Self-contained Irradiators

As noted previously, the form of the cesium-137 in self-contained irradiators is the same as the source in the Goiânia radiation dispersal accident: radioactive cesium chloride powder. The Goiânia accident showed that the cesium chloride salt pellets are easily dispersed if the source container is breached. Another problem with these and other nondefense high-activity cesium-137 sources is that they currently have no permanent disposal pathway in the United States. As noted in Chapter 2, they are considered "Greater than Class C" low-level radioactive waste, and the United States has not yet established a permanent disposal facility for such waste. The National Nuclear Security Administration's Offsite Source Recovery Project (OSRP) recovers unwanted and abandoned sources, so that licensees are not stuck with their sources indefinitely. The OSRP is discussed further in Chapters 2 and 10.

Teletherapy and Gamma Knife®

The U.S. NRC reports that there are over 240 cobalt-60 teletherapy units in the United States, although most are not now used for radiotherapy. The sources are compact and intense gamma emitters, and so would primarily be of interest for an RED, although they contain thousands of small cobalt-60 pellets which could be dispersed (see the discussion of the accident in Juarez, Mexico, later in this chapter). The sources in a Gamma Knife® are similar to those used in teletherapy, but the individual sources in a Gamma Knife® are much lower activity than the teletherapy sources (roughly 10 TBq versus perhaps 550 TBq). Each source is held in the Gamma Knife® container in such a way that retrieval of the 201 sources is a slow process requiring specialized tools. Just like in any other use of high-activity radionuclide sources, the security involved with the transport of fresh cobalt-60 sources from the manufacturer to the customer is of some concern.

---

[5] A recent National Research Council report (2006b) recommends "an independent examination of the security of spent fuel and high-level waste transportation .... provide an integrated evaluation of the threat environment, the response of packages to credible malevolent acts, and operational security requirements for protecting spent fuel and high-level waste while in transport." The security concerns about shipment of radionuclide radiation sources are similar to those about shipment of radioactive waste.

RTGs

The former Soviet Union produced approximately 1,000 RTGs for supplying remote power to navigational beacons and lighthouses. Many of these were abandoned after the breakup of the Soviet Union. In addition to concerns about inadvertent exposures, officials in the United States and Russia are concerned that an RTG source might be used in an RDD. The U.S. government and other international partners are currently helping Russia and other former Soviet countries locate and recover abandoned RTGs (see National Research Council, 2007), improve security for RTGs still in use, and replace some with alternative technologies (IAEA, 2005b). The RTGs in the United States are housed on government facilities that are required to have robust security for other reasons. The committee did not explore replacement technologies for RTGs because they are not commercial sources and because the former Soviet sources are not in use in the United States.

Well Logging Neutron Sources

Well logging sources are used wherever boreholes are found or are being drilled, especially in oil-rich areas of the United States. There are thousands of nuclear logging tools and they are transported by truck from the oil-field service companies to the job sites. The trucks, the devices that use the sources, and the sources themselves are expensive and are provided with some security because of their cost.

Industrial Radiography

Some (perhaps most) of the thousands of radiography devices are portable (i.e., can be carried by a person) and are used out in the field, making them more vulnerable to theft. Mobile radiography units are heavier devices mounted on wheels or placed on dollies for mobility. Still others are neither portable nor mobile. The U.S. NRC's Nuclear Material Event Database lists a number of incidents in which radiography devices containing radionuclide radiation sources were lost or stolen. Balancing this is the fact that iridium-192 has a short half-life and is not readily dispersible.

Internalizing the Costs of Security Risks

As noted in Chapter 1, not all social costs of radiation source use are borne by the users of the radiation sources. One option for implementing replacements that is discussed in Chapter 10 of this report is to make the users bear more of those costs, that is, to "internalize" the costs. With respect to security, the users already bear at least some of these costs.

After 2001, the U.S. NRC imposed enhanced security requirements on its materials licensees: Compensatory Measures for panoramic and underwater irradiators, Additional Security Measures for its manufacturers and distributors, and Increased Controls for licensees with Category 1 and 2 devices and sources. The commission (along with the Department of Transportation, with whom the commission shares regulatory authority) also imposed more stringent requirements on transportation of RAMQC. Compensatory Measures, Additional Security Measures, and Increased Controls include varying levels of access controls and alarms with response by security personnel coordinated with local law enforcement and coordination with companies shipping to the facility. The revised shipping requirements are different for Category 1 and Category 2 quantities, and include maintaining constant control and/or surveillance during transit and physical controls to serve as barriers to unauthorized removal. The details of the security provisions at any particular site or shipment are not public, and the committee did not review the adequacy of the security. The U.S. NRC indicated that it has carried out inspections for large, panoramic irradiators and manufacturers and distributors and will complete inspections related to the transportation requirements in 2007. Inspections for other licensees who possess Category 1 and 2 sources are scheduled for completion in the summer of 2009. The Agreement States have issued legally binding requirements equivalent to the U.S. NRC's Orders for Increased Controls. The U.S. NRC told the committee that it plans to promulgate regulations that codify the enhanced security requirements, taking into account lessons learned from their implementation by the U.S. NRC and the Agreement States.

Although the committee has not examined the U.S. NRC security requirements in detail, it is aware that their application is based on the IAEA categorization system, specifically applied to licensees possessing IAEA Category 1 and 2 sources/devices. As part of their analysis to identify risk-significant sources and quantities, the U.S. NRC and the Department of Energy (DOE) considered the potential for the material to cause deterministic health effects and contamination of an area greater than 0.5 km^2 in excess of the Environmental Protection Agency's intermediate-phase protective action guide. The latter was a threshold criterion that factored in the potential for contamination (DOE/U.S. NRC, 2003). The U.S. NRC concluded that the results of its own assessment were not significantly different from those found in the IAEA system, and so adopted the IAEA categorization system.

The IAEA categorization, as discussed in Chapter 1, is based on defining a "D value" (D for dangerous level) for each radionuclide. Category 1 sources are those that exceed the D value by a factor of 1,000, while Category 2 sources are greater that 10 but less than 1,000 times the D value. Table 3-1 lists the D values and IAEA Category 1 and 2 thresholds for the radionuclides of interest. A low D value indicates a highly hazardous source and a correspondingly low threshold for Additional Security Measures. The key characteristic that yields a low D value is highly energetic decay. Cobalt-60 emits two high-energy gamma rays with each decay, compared with approximately one moderate energy gamma ray for each decay of cesium-137, and so the D value for cobalt-60 is more than a factor of three lower than the D value for cesium-137. Because the IAEA categorization is based on deterministic health effects and safety concerns, it factors in considerations relevant for RED risks and scenarios. It does not, however, account well for RDD risks, which are dominated by area-denial aspects of dispersed radioactive material. Although the DOE/U.S. NRC analysis to identify sources and quantities of concern did include area contamination, it was a single threshold criterion. That is, it could have affected which radionuclide radiation sources are in Category 2, but there would have been no distinction, in terms of the potential for contamination, between Category 2 and Category 1.

TABLE 3-1 IAEA D Values and Category 1 and 2 Thresholds

Radionuclide	**IAEA D Value** (TBq) [Ci]	**IAEA Category 2 (10 x D)** (TBq) [Ci]	**IAEA Category 1 (1,000 x D)** (TBq) [Ci]
Am-241	0.06 [1.6]	0.6 [16]	60 [1600]
Co-60	0.03 [0.81]	0.3 [8.1]	30 [810]
Cs-137	0.1 [2.7]	1.0 [27]	100 [2700]
Ir-192	0.08 [2.2]	0.8 [22]	80 [2200]
Pu-238	0.06 [1.6]	0.6 [16]	60 1600]
Sr-90	1.0 [27]	10 [270]	1,000 [27,000]

SOURCE: Courtesy of IAEA (2005a).

Further, it did not account for differences in cleanup of the contamination or any other factors that might contribute to the economic consequences of an attack. Ultimately, the DOE/U.S. NRC 0.5-km^2 criterion may have affected the quantities that define the source categories for a few radionuclides,[6] but had no impact on others. To evaluate the adequacy of its security measures, the U.S. NRC carried out security assessments for nuclear materials facilities considering several attractiveness factors in assessing the threat but only deterministic health effects from radiation exposures as the consequences of interest. The U.S. NRC staff characterized this as the commission's first step in reevaluating security needs for materials licensees.

A comprehensive and quantitative examination of the area-denial risks of radiation sources is beyond the scope of this study. However, the key characteristics that make radiation sources hazardous with respect to area-denial RDDs are summarized in Figure 3-2. They are the availability of large radiation sources, the dispersibility of the sources (see Sidebar 3-1), the persistence of the radioactive material once is has been dispersed (this reflects both radioactive decay and the tendency of the material to bind to other materials in the environment), and the potential to cause harm to human health and the environment, which affects the long-term cleanup goal. Several publications (see e.g., U.S. NRC, 2007c; Harper et al., 2006; NCRP, 2006; Argonne, 2005; Van Tuyle et al., 2003; NCRP, 2001) describe features of some or all of the most common high-activity radionuclides with respect to some of these key RDD characteristics. Comparing the radionuclides that emerge from these characteristics, a different ordering of hazards emerges from those shown in Table 3-1 for deterministic health effects.

The committee does not advocate a change to the categorization system that is already in place. The IAEA categorization system is being used for multiple purposes by both the international community and the U.S. NRC, spanning regulatory, safety, and security guidance. The system has been used to establish new regulations on the import/export of radioactive material. It is also used both domestically and internationally to help prioritize the recovery of orphaned and unwanted radiation sources. And, as noted earlier, the U.S. NRC and the Agreement States have applied the Increased Controls orders (enhanced security measures) to licensees in possession of IAEA Category 2 and 1 quantities of radioactive material.

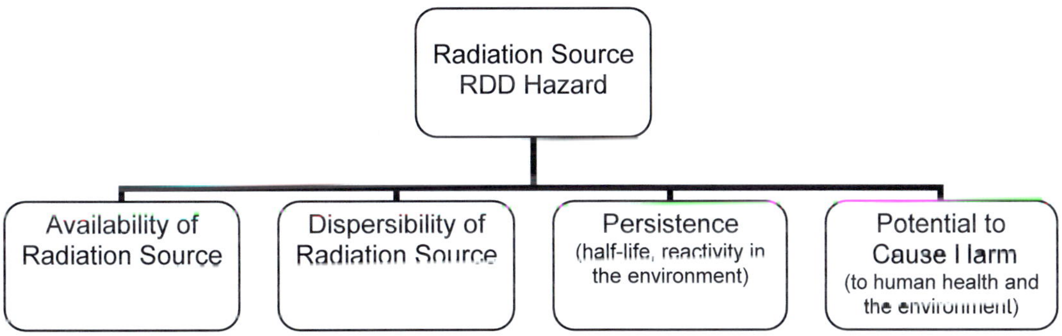

FIGURE 3-2 Key characteristics that make radiation sources hazardous with respect to area-denial RDDs.
SOURCE: Image provided by the committee.

---

[6] To the committee's knowledge, there is no documentation that these calculations affected the IAEA categories. The U.S. NRC informed the committee that information the U.S. agencies shared did have effects on the category thresholds for a few radionuclides, and a person involved in the IAEA effort confirmed that conversations with the U.S. agencies led to rethinking of some thresholds.

SIDEBAR 3-1
FRAGMENTATION

Dispersal of radioactive materials in an explosion has emerged as a possible hazard scenario. Fragmentation during an explosion or as a result of impact has been the subject of extensive studies since the Second World War. A distinction is made between brittle materials, such as ceramics and glasses, and ductile materials, such as most metals. For brittle solids, fracture mechanics models predict that the size, *d*, of dust fragments is related to the fracture toughness for high-speed fracture, K_{cr}, of the solid:

$$d = 2.9\left[\frac{K_{cr}}{\rho\, c_0\, \acute{\varepsilon}_0}\right]^{2/3} \qquad 3\text{-}3,$$

where ρ is the density, c_0 is the elastic wave speed and $\acute{\varepsilon}_0$ is the deformation rate (see Grady (1982) and quoted by Freund (1998)). The fracture toughness is an intrinsic property of a material and its processing. Most metals undergo plastic deformation before fracture and have considerably larger fracture toughness than brittle solids. Their ductile behavior makes fragmentation into small dust particles much less likely and instead they break or tear into larger pieces.

Iridium is a face-centered cubic metal but, unlike the majority of metals with that structure, such as gold and aluminum (which are ductile and show considerable malleability), iridium exhibits little ductility and fractures in a brittle manner. Americium-beryllium sources, such as those used in oil-well logging, typically are formed by sintering mixtures of americium oxide and beryllium powders to form a pellet which is then either diffusion bonded to a metal strip or sealed in a welded stainless steel container.

The committee does, however, conclude that consequences other than deterministic health effects, especially the consequences of area-denial RDDs, should be factored into decisions about security for radiation sources. The committee judges that area-denial RDDs have the greatest potential consequences among the kinds of possible attacks with at least some of the high-activity radiation sources, and may pose the greatest risks, as well.

Returning to Figure 3-1, for each element of probability and consequence, there are measures that can be taken to either reduce the probability or mitigate the consequences. On the probability side, for example, the probability of acquiring an RDD-significant quantity of radioactive material can be reduced by increasing the security of such materials, tightening the regulatory controls on their use, or by reducing the overall quantity of radioactive materials in use.

Similarly, by understanding the relative difficulty of RDD manufacture and RDD effectiveness based on the radionuclide used, the government can better prioritize those radionuclides based on which is most hazardous in RDD scenarios and which poses the greatest RDD risks. This would enable the government to apply greater security and control over those radionuclides. Preventing an attack is clearly preferable to dealing with the consequences, but addressing the consequence side, which involves measures that mitigate the consequences of an RDD attack should one occur, may be just as important. Another National Research Council study (2002) found that better public awareness and education about the true risks of an RDD attack and a clear, well-planned response would greatly help in mitigating the psychological impact on the public.

Taken as a whole, the mix of government countermeasures represents a layered defense against RDD attack. No single layer can be perfect because gaps will always exist and determined terrorists can take the time to test the defenses to find weak points. The defensive layers, therefore, must be examined as a system and improvements in defenses must come from identifying and addressing system weaknesses. Thus, even if one cannot numerically calculate the risk of an RDD attack, this discussion shows that the concept of risk is helpful in organizing our thinking about the problem. It is also useful in understanding how and where defensive countermeasures can be taken and which part of the RDD risk equation is being affected.

Radioactive Cesium Chloride

This National Research Council study is devoted to examining the options for replacing high-risk radiation sources with alternatives that pose lower risks of malevolent use. In terms of Figure 3-1, this replacement would impact the Source Material Acquisition box, thus reducing the probability that terrorists could acquire a radiation source that would pose a high risk if used in an attack. When evaluating the potential harm that could be caused by different radionuclides in radiation sources, radioactive cesium chloride sources emerge as a major concern.

Cesium chloride is soluble and highly dispersible (other forms of cesium-137 mentioned in Chapter 2 are not as dispersible). It emits penetrating radiation and so it cannot be easily shielded if dispersed. Devices containing sizable quantities of this material are used across the United States, most commonly in facilities that are located in cities, large and small, which are potentially attractive targets, and the number of these sources appears to be increasing. All of these factors contribute to making radioactive cesium chloride such a concern to the committee.

This concern is exacerbated by the lack of an avenue for permanent disposal of high-activity cesium radiation sources, which increases the likelihood that unwanted cesium radiation sources will remain in unplanned storage where they are potentially more vulnerable to theft. The alternative available to owners of these sources is to obtain the services of the Conference of Radiation Control Program Directors (CRCPD) source disposition program or the OSRP, which have been subject to budget uncertainties.

The IAEA efforts have identified the same characteristics and concerns. In the findings from a 2003 IAEA conference, the group encouraged IAEA (2003a):

- to formulate and implement national plans for the management of radioactive sources throughout their life cycle;
- to develop, to the extent practical, standards for the design of sealed sources and associated devices that are less suitable for malevolent uses (e.g., alternative technologies, less dispersible forms of radioactive sources);
- to establish arrangements for the safe and secure disposal of disused high-risk radioactive sources, including the development of disposal facilities.

In the same year, the IAEA guidance on security of radioactive sources (IAEA, 2003b) identified the typical form of cesium-137 as "radioactive material that could be easily dispersed via an explosion or otherwise destroying the source."

The IAEA report on the Categorization of Radiation Sources (IAEA, 2005a) also presents an overview of radiation source applications. Figure 3-3 presents an abridged summary of the IAEA data. That figure displays the various applications on the vertical axis, and the activity range per application on the horizontal axis. The horizontal bars represent the range of activity levels for each application while the black vertical line within each bar delineates the typical activity used. The most common radionuclide used for each application is also listed next to each bar (a number of different radionuclides are used for most applications, but the figure shows only the most common nuclide). Note that these applications cover a very wide range of activities (12 orders of magnitude), from tens-of-kBq (microcurie) smoke detectors to hundred-thousand-TBq (megacurie) sterilization irradiators. The thick line running through the applications represents the radionuclide specific threshold levels for IAEA Categories 1 and 2.

The interagency Radiation Source Protection and Security Task Force formed at the direction of Congress also highlighted cesium chloride as deserving special attention (U.S. NRC, 2006a):

> A specific area of concern is the widespread use of cesium chloride (CsCl) in a highly dispersible form in certain devices. The Task Force recommends that high priority be given to conducting a study within 2 years to assess the feasibility of phasing out the use of CsCl in a highly dispersible form. This study should include consideration of the availability of alternative technologies for the scope of current uses, safe and secure disposal of existing material, and international safety and security implications. Any plan to phase out these sources should involve industry and consider not only alternatives for uses of these materials, but also how to compensate owners of these sources so that they do not find their way into environments where less rigorous controls are in place.

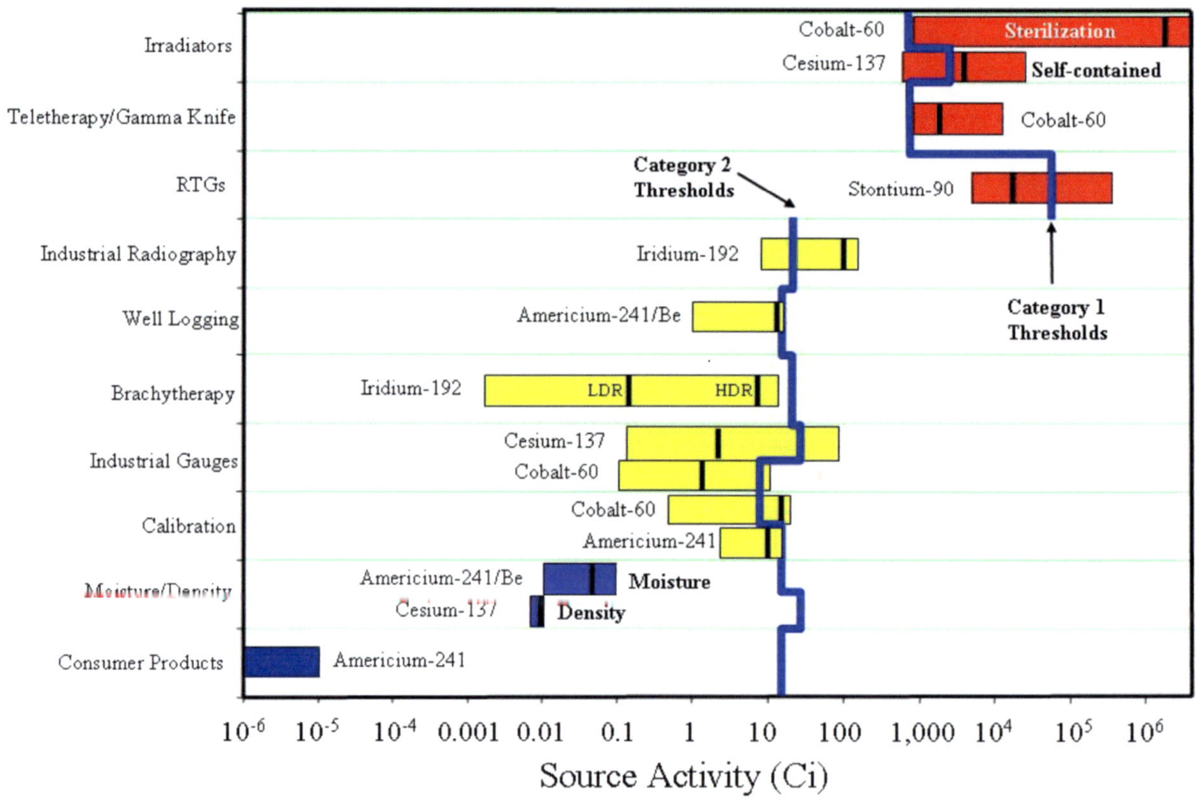

FIGURE 3-3 Radiation source applications, radionuclides, and activity ranges. NOTE: 1 Ci = 0.037 TBq. SOURCE: Modified from IAEA (2005a).

The consequences of dispersal of cesium-137 in the environment can be understood more clearly by examining three accidents involving this radionuclide.[7] These events took place in Chornobyl,[8] Ukraine; Goiânia, Brazil; and Decatur, Georgia, in the United States. In addition, an accident involving dispersal of cobalt-60 in Juarez, Mexico, illustrates some of the differences between dispersal of discrete radionuclide radiation sources and finely divided radioactive material. Each of these is discussed below. It should be noted that the scale and manner of dispersal of radioactive material by the reactor accident is different from an RDD. Chornobyl dispersed orders of magnitude more cesium-137 than is contained in any device in the United States, but lessons can still be learned from the accident.

Relevant Lessons from the Chornobyl Accident

The Chornobyl reactor accident of April 1986 (IAEA, 1991) resulted in dispersal of radioactive material worse than any other. A reactor test gone awry led to a sudden tremendous power excursion, causing an explosive breach of the reactor vessel and a subsequent graphite fire that released a plume containing over 3.7 million TBq (100 MCi) of radioactive material. Several days passed before some of the exposed population was made aware of the accident and evacuated. The first responder community of firefighters and other rescue teams suffered the immediate health consequences of Chornobyl—over 30 of them died from acute radiation exposure. A delayed health impact has been seen in increased incidence of cancer. The children exposed to some of the roughly 1.8 million TBq (approximately 48 MCi) of iodine-131 released in the accident (IAEA, 2006) were the most affected members of the population for delayed health effects. It is worth noting that iodine-131 is not available in Category 1 and 2 quantities for commercial use in either the United States or the world market.

As significant as these consequences are, the massive economic impact caused by the ground contamination from the cesium-137 released by the Chornobyl accident is its own national-scale disaster. Although many of the radioactive species released from a reactor accident are short-lived or are noble gasses that disperse to harmless levels in the atmosphere, two particular species persist as contamination. They are cesium-137 and strontium-90, both of which are long-lived (approximately 30-year half-lives) and have sufficient dose potential to pose continued risks.[9] The Chornobyl accident released roughly 70,000 TBq (2 MCi) of cesium-137 and approximately 11,000 TBq (300,000 Ci) of strontium-90. Thus, the cesium-137 was the dominant radionuclide for ground contamination. The cesium-137 surface contamination was measured, using standard radiation detection equipment. It is presented graphically in terms of the quantity of cesium-137 per square kilometer of surface area, TBq/km^2 (Ci/km^2) in Figure 3-4, which shows the cesium-137 ground contamination around the Chornobyl site.

After much debate and confusion, the Soviet government finally settled on criteria for relocating the populations living on contaminated ground. The upper limit was 1.5 TBq/km^2 (40

---

[7] The National Council on Radiation Protection and Measurements recently issued an informative report titled *Cesium-137 in the Environment: Radioecology and Approaches to Assessment and Management* (2006), which describes cesium-137 contamination in the environment due to releases at sites in the United States, the Ukraine, and Brazil. The report describes near-term countermeasures and long-term cleanup strategies as well.

[8] The name of this city and nuclear power plant are commonly spelled Chernobyl in English as a transliteration of the Russian pronunciation. The committee uses here a transliteration of the Ukrainian spelling (Чорнобиль) because the power plant is located in the Ukraine.

[9] The cesium-137 deposited as fallout during the accident was not in the form of cesium chloride, but once deposited in a moist environment, the cesium behavior is similar.

Ci/km^2). At this level of contamination, the area was confiscated and the population forcibly removed. No attempt was made to clean up these areas, called confiscated zones (the area illustrated with a mesh of lines in Figure 3-4). The confiscated zones amounted to approximately 3,000 km^2, and a population of over 300,000 was relocated. The permanent control zone involved contamination levels from 0.5 to 1.5 TBq/km^2 (15 to 40 Ci/km^2). In this zone, the inhabitants were given the option to be relocated or to stay and receive financial compensation.

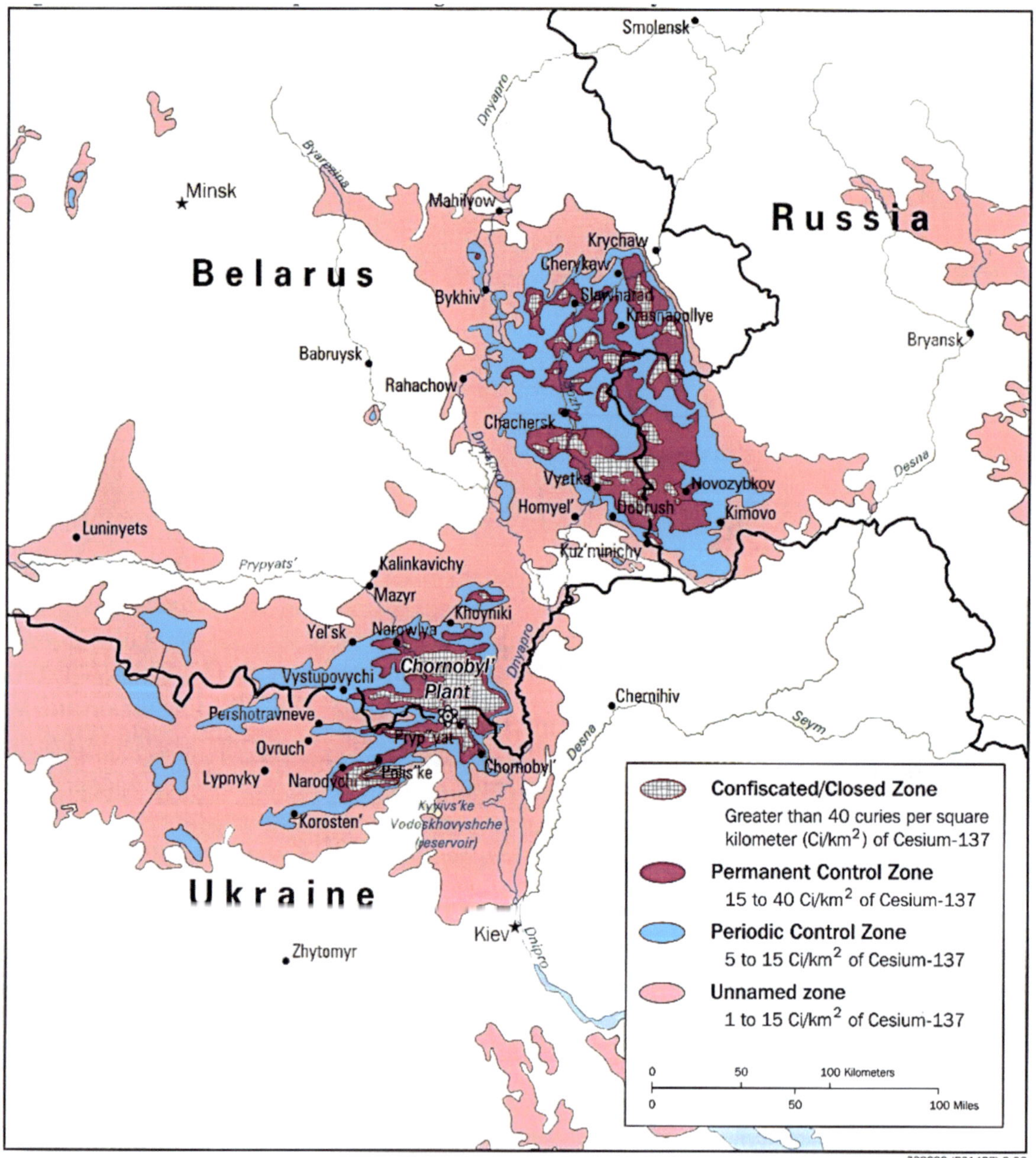

739928 (R01428) 9-96

FIGURE 3-4 Cesium-137 ground contamination zones from the Chornobyl accident. SOURCE: Image adapted for black-and-white reproduction from De Cort et al. (1998).

From 1986 to 1989, decontamination measures were taken at tens of thousands of residences and public buildings, as well as more than 1,000 farms. Contamination levels were reduced by factors ranging from 10 to 100.

In addition to the huge economic costs of cleanup and relocation, the psychological impact on the affected populations of the Ukraine and Belarus were also very significant. According to the 2005 UN review of the Chornobyl incident (WHO, 2005):

- Persistent myths and misperceptions about the threat of radiation have resulted in "paralyzing fatalism" among residents of affected areas.
- Alongside radiation-induced deaths and diseases, the report labels the mental health impact of Chornobyl as "the largest public health problem created by the accident" and partially attributes this damaging psychological impact to a lack of accurate information. These problems manifest as negative self-assessments of health, belief in a shortened life expectancy, lack of initiative, and dependency on assistance from the state.

Relevant Lessons from Goiânia

Another serious radiation dispersal accident again involving cesium-137 occurred within two years of the Chornobyl disaster, in Goiânia, Brazil, in September 1987 (IAEA, 1988). An abandoned teletherapy device (see Figure 3-5) was found by scrap metal scavengers. They managed to remove the heavy metal shielding subassembly from the machine (which also contained the sealed source) and, while trying to remove the source holder from the shielding, punctured the 1-mm-thick window of the source capsule with a screwdriver and removed some of the radioactive contents. The scrap was sold to a local junkyard, which then had possession of the source and the remainder of its contents, now exposed to the environment.

The cesium-137 source emitted a blue glow and the junkyard owner, believing it to be valuable and possibly supernatural, removed pieces of the source and distributed it to friends and family. The cesium-137 was thus unintentionally dispersed throughout various parts of Goiânia, simply by human handling and normal traffic of humans and animals about the city. The easy dispersal was facilitated by the form of cesium-137 used in these and other high-activity devices in use throughout the world, then and now. As described in Chapter 2, the cesium-137 is in the form of radioactive cesium chloride. Figure 3-5 shows a diagram of the teletherapy device and a picture of the kind of sealed source holder used in the device. The radiation source contained 51 TBq (1375 Ci) of cesium-137. As can be seen from Figure 3-5, the source is small. It was approximately 60 grams of cesium chloride, a quantity that would easily fit inside a typical salt shaker.

This quantity of cesium-137, if uniformly distributed, could contaminate an area of approximately 35 km^2, using the relocation/confiscation zone criterion of 1.5 TBq/km^2 (40 Ci/km^2) from the Chornobyl accident. Fortunately, creating a uniform dispersal is difficult, and in Goiânia the dispersal covered a region of only 1 km^2. Nevertheless, this small amount of cesium-137 created a huge cleanup problem for the city, resulting in the generation of over 40 tons of radioactive material for disposal.

Many RDD relevant lessons can be gleaned from the consequences of the Goiânia accident: First, there were only four initial deaths,[10] and primarily among those who actually

[10] Over 112,000 people were surveyed for contamination; 249 people were contaminated and of those, 129 had both internal and external contamination. Some 50 people showed signs of whole-body irradiation, and radiation injuries were observed on 28 of them. Fourteen of the 50 had damage to their bone marrow and required intensive care. Four of these people died within 2 months of the accident.

handled and accidentally ingested cesium-137. Second, psychosocial consequences were considerable; citizens of Goiânia were shunned by the rest of the country and many Goiânians who received no radiation exposure presented with psychosomatic symptoms of radiation sickness. Third, cesium-137 salt is readily dispersible and very active and mobile in the environment. Once on the ground the cesium-137 salt went into solution with the ground moisture. When the ground moisture evaporated in daytime heat, cesium-137 dust particles became airborne, thus enlarging the dispersal area. The large quantities of radioactive waste produced in Goiânia were the result of cesium-137 chemically bonding to standard building materials (such as the tile roofs shown in Figure 3-5). It was not economically feasible to remove the cesium-137 contamination from these surfaces, resulting in the demolition of several contaminated structures.

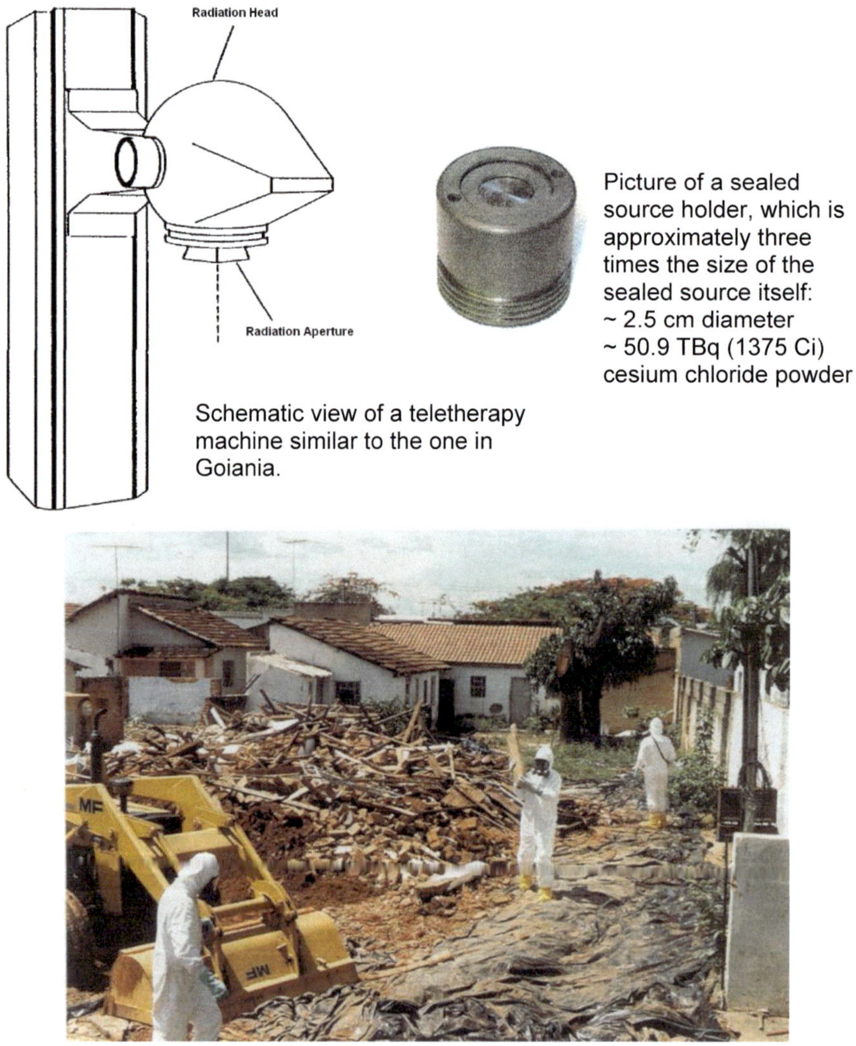

FIGURE 3-5 The Goiânia radiation dispersal accident. SOURCE: Image courtesy of IAEA (1988).

Many of the people who received high doses (more than 1 Gy or 100 rad) have persistent medical conditions resulting from the exposure. In May 1994, another person who was severely exposed (7.0 Gy) died of chronic liver failure, which is likely a result of the dose received (see Brandao-Mello et al., 2000).

The cleanup criterion used for exterior surfaces in Goiânia was 0.37 TBq/km^2 (10 Ci/km^2). This corresponds to an annual projected dose to inhabitants (using the standard dose conversion factors described previously) of approximately 5 mSv/yr (500 mrem/yr). This projected dose is roughly comparable to the dose received by the general population from natural background radiation, about 3 to 4 mSv/yr, depending on location. Note that no permanent relocations were performed at Goiânia, all the contaminated zones underwent cleanup, and this threshold for initiating decontamination operations (10 Ci/km^2) was lower by a factor of four than the Soviet relocation criterion of 40 Ci/km^2.

Decatur, Georgia and the Qualified Ban on Cesium Chloride Sources in Panoramic Irradiators

In 1986, Radiation Sterilizers, Inc. (RSI), requested permission to use 252 capsules containing a total of 428,000 TBq (12.3 MCi) of radioactive cesium chloride from DOE in a wet-storage panoramic irradiator. The U.S. NRC had announced the previous year that it would accept applications for use of the DOE capsules. The cesium-137 sources provided by DOE were much larger than typical cesium chloride radiation sources sold today. DOE had Sandia National Labortories carry out tests with the capsules in wet-load, dry-storage, dry-irradiator mode. This testing did not demonstrate the capsules' performance under conditions that cycle the temperature, as would happen in a wet-storage, dry irradiator, so another test campaign was begun at a different RSI irradiator in Westerville, Ohio. The U.S. NRC approved the use in other irradiators before that campaign was completed, and the Decatur facility received permission to use the capsules (DOE, 1990; Setser, 1990). In June 1988, about 300 GBq (8 Ci) of cesium-137 leaked into cooling water from a source containing over 1,850 TBq (50,000 Ci) of cesium-137. The leak apparently was caused by stress from thermal expansion of the bulk cesium chloride, which had melted and relocated to form a block at one end of the source capsule.

After the accident, three task forces were formed to investigate the causes and lessons learned: One team was formed by the governor of Georgia, another by DOE, and a third by the CRCPD. The Georgia Task Force, in commenting on the future use of the capsules involved in the incident, quoted a comment from the IAEA report on the Goiânia accident and drew the following conclusion (Setser, 1990):

> '[T]he physical and chemical properties of radioactive sources are very important in relation to radiological accidents. They should be taken into account in the licensing of the manufacture of such sources, in view of the potential influence on these properties on the consequences of accidents with the use or misuse of sources.' This is not an issue to be taken lightly by DOE or NRC. This issue needs to be fully resolved to the satisfaction of all cognizant regulatory agencies involved.

The revised U.S. NRC regulation for panoramic irradiators requires that the sources have a certificate of registration issued under 10 CFR § 32.210, be doubly encapsulated, and use radioactive material that is as nondispersible as practical and that is as insoluble as practical if the source is used in a wet-source-storage or wet-source-change irradiator (10 CFR Part 36). The regulation concerns only panoramic irradiators and underwater irradiators, not the dry-storage self-contained irradiators discussed elsewhere in this report.

In February 2001, the U.S. NRC denied a source certificate to the company GrayStar for a GS-42 sealed source containing 1,900 TBq (51,500) Ci of cesium-137 chloride in "caked powder" form. GrayStar designed a dry-storage irradiator for food irradiation using 64 of these doubly encapsulated sources (total of approximately 122,000 TBq [3.3 MCi]). The denial was based on the dispersibility and solubility of these sources and the requirement in 10 CFR § 32.210 "to provide reasonable assurance that the radiation safety properties of the source or device are adequate to protect health and minimize danger to life and property" (U.S. NRC, 2001).

GrayStar argued that cesium dispersed in glass by vitrification would require a greater amount of material to achieve the same irradiation levels as cesium-137 chloride and that the complexity of producing compounds other than cesium-137 chloride would cause "major difficulties and complexities in hot cell operations for source preparation." The U.S. NRC Atomic Safety and Licensing Board rejected all of these arguments (U.S. NRC, 2001):

> The Commission determined that the safety hazards associated with leaks of dispersible cesium chloride, even though the leaks were infrequent, justified restricting its use....The Staff argues that the longer half-life and decay time of cesium-137, combined with its dispersibility, could actually present an increased risk in comparison with cobalt-60 if a leak or other safety problem occurred, including the possible increased risks associated with a proliferation of smaller (non-cobalt) irradiators "in the vicinity of food processors, whose personnel have no previous training or experience with radiation safety." ...The [NRC] staff contends that ... cesium-137 chloride powder – even in its "caked" form – is dispersible, not only in water but also in air, by physical forces such as air turbulence, physical contact, fire or explosion, should there be a leak in any of the source capsules.... The Staff asserts that the Commission's nondispersibility requirement "reflects its general defense-in-depth philosophy, in that it assumes sealed sources will leak, and guards against the consequences caused by the spread of radioactive material *after a breach occurs.*

Relevant Lessons from the Accident in Juarez, Mexico[11]

In late 1983, two men working for the Centro Médico in Juarez, Mexico, hauled some material and equipment taken from the hospital's warehouse to a junkyard, Jonke Fénix, across town. Among the equipment dismantled in the warehouse was a Picker 3000 cobalt-60 teletherapy device purchased used from a U.S. company. The 20-year-old device was loaded with a new 107-TBq (2,885-Ci) cobalt-60 source in September 1969. One of the handymen transporting the material from the warehouse took out the radionuclide radiation source capsule and, not knowing what it was, pried it open, and ultimately sold the remains of the capsule to the junkyard. Some of the 6,010 cobalt-60 pellets that the radiation source contained were scattered in the pickup truck and on the road, but most were spilled at a few locations in the junkyard where the loading magnet picked many of them up and intermingled them with the scrap metal.[12] The junkyard sent scrap metal carrying the pellets to two steel foundries in Chihuahua, which melted them and produced contaminated steel. Correcting for decay to the time of the accident, the cobalt-60 source contained 16.5 TBq (445 Ci), or about 2.7 GBq (74 mCi) per pellet.

No one was aware of the contamination until an investigation was initiated by sheer happenstance: A truck carrying contaminated steel rebar took a wrong turn and passed over a radiation detector in Los Alamos National Laboratory that triggered an alarm in 1984. That alarm led authorities in the United States and Mexico to trace the history of the steel. Hundreds of

---

[11] This discussion is taken from TCPA (1998) and Marshal (1984).
[12] Cobalt is a ferromagnetic metal, like iron and nickel.

people received significant doses of radiation, including at least four who received between 3 and 4.5 Sv (300 to 450 rem) as whole-body doses. (Based on statistics, one would expect that half of the people in a group receiving doses around 4 Sv would die from the radiation.) About 4,000 tons of steel was contaminated with about 11 TBq (300 Ci) of the cobalt-60, including 600 tons sent to 23 states in the United States. Some 3,400 tons of steel stayed in Mexico and 109 houses built with contaminated rebar were demolished on orders from Mexican health officials. Using a borrowed helicopter and a radiation detector, officials checked the roads between the foundries and the junkyard, and found 22 contaminated sites, including eight pellets that were embedded in the road.

The scale of this accident is larger than one would expect from dispersal of discrete pellets of cobalt-60 because the actions of industrial equipment at the junkyard and the foundries dispersed them more widely and finely. The fact that a survey could be done by helicopter and individual pellets could still be recovered illustrates why scattering discrete pellets of cobalt-60, even in large numbers, imposes less of a cleanup burden than dispersal of radionuclide radiation sources with more finely divided material. Because of this incident and several others, foundries in the United States now have radiation detectors, as do the U.S. land border crossings and major shipping harbors.

FINDINGS AND RECOMMENDATIONS

Finding: The U.S. NRC ranks the hazards of radiation sources primarily based on the potential for deterministic health effects (especially death and severe bodily harm) from direct exposure to the radiation emitted by the bare (unshielded) sources. The U.S. NRC's analyses that support the commission's security requirements for nuclear materials licensees are based only on these potential consequences.

The U.S. NRC has ranked radiation sources in terms of hazard using the IAEA system of five source categories, determining that the Category 1 and 2 sources are "high-risk sources." The IAEA analyses supporting its source categorization system consider only deterministic health effects (such as early fatalities) from direct exposure to ionizing radiation from the unshielded source under different exposure scenarios. The initial DOE/U.S. NRC analysis used the same consequences and added in a contamination criterion that falls short of reflecting the area-denial or economic consequences of a dispersal attack. The U.S. NRC also carried out security analyses of each type of facility licensed to use Category 1 and 2 sources, but these analyses were confined to examining the potential for deterministic health effects caused by attacks involving the Category 1 and 2 sources. The U.S. NRC staff told the committee that this was seen as a first step, and that the commission was considering whether to include other factors.

Finding: Factors other than the potential to cause deterministic health effects are important when evaluating hazards from radiation sources, especially the potential to cause contamination of large areas resulting in economic and social disruption (area denial).

A radiological incident (an accident or especially an attack) could have its most long-lasting and far-reaching effects as a result of contamination of land, buildings, and infrastructure in densely populated regions, partially or completely disabling those assets for human use for long periods of time. This is illustrated by the radiotherapy source incident in 1987 in Goiânia,

Brazil, and the Chornobyl nuclear reactor accident in the Ukraine. Although an event like the Chornobyl reactor fire is not possible with radiation sources and the scale of the contamination from an incident with radiation sources would inherently be smaller, that 1986 accident showed that radioactive contamination can create sizeable areas that are deemed uninhabitable. The economic and social disruptions caused by such incidents can be difficult to quantify, but they are critical to understanding the scope of the impact beyond the fatalities and severe bodily injuries caused by these events.

Recommendation: For prioritizing efforts to reduce risks from malicious use of radiation sources, the U.S. NRC should consider radiation sources' potential to cause contamination of large areas resulting in economic and social disruption (area denial) to determine what, if any, additional security measures are needed.

Having taken an essential first step in considering deterministic health effects from possible radiation exposure from an incident involving radiation sources, the U.S. NRC should now include economic and social disruption in its risk analyses of radiation sources. These impacts can vary significantly depending on the scenarios considered, but that variability does not make them less important. Further, even with such variability, certain factors emerge as important in other analyses of these issues (e.g., Van Tuyle et al., 2003). In carrying out its analyses, the U.S. NRC should not confine itself to the numeric source-activity cutoffs defining the lower limits for Category 1 and 2 sources because the source categorization system itself is based on deterministic health effects. For example, many self-contained irradiators are Category 2 devices, but are near the Category 1 threshold and most americium-beryllium well logging sources have activities near but below the Category 2 limit. Review may show that each set of devices should be regulated similarly.

After 2001, the U.S. NRC imposed enhanced security requirements on its materials licensees: Compensatory Measures for panoramic irradiators, Additional Security Measures for its manufacturers and distributors, and Increased Controls for licensees with Category 1 and 2 devices and sources. Compensatory Measures include fairly robust access controls and alarms with response by armed security personnel, along with other measures. Increased Controls include access controls and alarms with response by security personnel, and other measures. After review of the risks associated with some sources and devices, considering more fully the potential for contamination from an attack, the U.S. NRC might conclude that more stringent measures are needed at facilities licensed for some Category 1 and 2 sources and devices. The committee did not examine these security matters in detail and so cannot prejudge the outcome of such analyses. The committee does note, however, that such measures could improve the security of the devices and create a disincentive for owning them.

Finding: Because of its dispersibility, solubility, penetrating radiation, source activity, and presence across the United States in facilities such as hospitals, blood banks, and universities, many of which are located in large population centers, radioactive cesium chloride is a greater concern than other Category 1 and 2 sources for some attack scenarios. This concern is exacerbated by the lack of an avenue for permanent disposal of high-activity cesium radiation sources, which can result in disused cesium sources sitting in licensees' storage facilities. As such, these sources pose unique risks.

Radioactive cesium chloride sources are in the form of a steel-encapsulated, compressed powder. The salt is highly dispersible and water soluble. There are approximately 1,300 high-activity cesium chloride devices (each with an activity of tens to hundreds of terabecquerels [hundreds to thousands of curies]) across the United States, nearly all of which are self-contained irradiators. The number of these devices and sources appears to be increasing.

Because it emits energetic gamma rays and its half-life is long enough that an irradiator does not need to be reloaded over the device's expected lifetime, cesium-137 has been the key component of self-contained irradiators for blood irradiation and research for many years. Cesium chloride is the least expensive and highest-specific-activity form of cesium-137 available today. Because of the nature of the applications that employ these irradiators, they are most commonly located in hospitals, blood banks, and universities, many of which are located in cities, large and small. The presence of these sizable sources in areas that are potentially attractive targets is a major factor making radioactive cesium chloride such a concern to the committee.

Finding: In view of the overall liabilities of radioactive cesium chloride, the committee judges that these sources should be replaced in the United States and, to the extent possible, elsewhere.

Finding: Discussions in Chapter 2 and 5 show that in most (and perhaps all) applications, radioactive cesium chloride can be replaced by (1) less hazardous forms of radioactive cesium, (2) radioactive cobalt, or (3) nonradionuclide alternatives. However, not all of these alternatives are available now, and all are currently more expensive than radioactive cesium chloride for the users.

Finding: Government action is required to implement replacement of radioactive cesium chloride sources because the alternatives cost more and the liabilities or social costs of the sources are not borne by the end users.

There is no indication that replacement of devices containing Catogory 1 and 2 radioactive cesium chloride sources with lower hazard alternatives will improve or worsen the performance of the devices in their standard and proper uses. The act of replacement incurs monetary costs and the replacements themselves currently cost more in most cases than the radioactive cesium chloride devices. All of these costs would be borne by the end users (paying more for the alternatives) and the current device manufacturers (depending on the price elasticity of demand and potential loss of sales). The benefits of replacement are in reducing the liabilities and social costs (including the costs associated with the risk of terrorist attacks and, in some cases, the full costs of disposal, discussed in Chapters 2 and 10), which are shared by the public rather than borne by the end users. Except in cases where the replacements prove to be cheaper, end users have little incentive to shift away from their current devices; and unless there is a demand for the alternatives, manufacturers are unlikely to invest in making the alternatives available. Government action can, however, provide the requirements or incentives to implement replacement.

Recommendation: In view of the overall liabilities of radioactive cesium chloride, the U.S. government should implement options for eliminating Category 1 and 2 cesium chloride sources from use in the United States and, to the extent possible, elsewhere. The committee suggests these options as the steps for implementation:

i. **Discontinue licensing of new cesium chloride irradiator sources.**
ii. **Put in place incentives for decommissioning existing sources.**
iii. **Prohibit the export of cesium chloride sources to other countries, except for purposes of disposal in an appropriately licensed facility.**

In Chapter 10, the committee offers several suggestions as its lead candidates for how to implement the replacement, and they are summarized here. First, to stop the addition of new Category 1 and 2 cesium chloride sources to the nation's inventory, the U.S. NRC should discontinue all new licensing and importation of these sources and devices. This includes import of new sources from other countries and recycling of sources from decommissioned devices. Second, many licensees may need incentives to decommission[13] their existing sources or devices because the devices still have use value. Indeed, there are now also disincentives to decommissioning beyond the loss of use, including the costs of decommissioning. Third, if the sources recovered from decommissioned devices (or the devices themselves) are simply sold outside the United States, then the sources are still potentially available for use in an attack on another country or even the United States. Therefore, disposition options are needed in the United States. These are discussed in more detail in Chapter 10.

The overall policy could make exceptions based on unique needs that cannot be met with alternative technologies, but the threshold for creating exceptions should be set high, similar to what the U.S. NRC has done for panoramic irradiators.

[13] U.S. NRC's technical definition of the term decommissioning applies mainly to facilities and involves removing licensed radioactive material to an extent that allows public release of the facility and termination of the license. The committee's usage here and throughout the report is slightly different: A decommissioned device is retired from service and sent to whatever disposition option is available (disposal or storage pending dismantlement and disposal), and the license for the device is terminated.

CHAPTER 4

ACCELERATOR AND DETECTOR TECHNOLOGIES

SUMMARY

Radiation can come from radionuclide decay,[1] atomic and nuclear reactions (e.g., alpha-*n* reactions), or from machine sources which are most commonly called accelerators, x-ray generators (including x-ray tubes), and neutron generators or neutron tubes. For the purposes of this report x-rays are electromagnetic radiation from machine sources, and gamma rays are electromagnetic radiation from radionuclide decay. Beta-minus particles from beta decay are electrons no different from the electrons in an e-beam from an accelerator. Similarly, neutrons are the same regardless of their source. Machines and radionuclide sources produce radiation with different energy distributions. Gamma decay emits radiation at discrete energies, although some radionuclides emit gamma rays at many different energies. X rays from accelerators and x-ray tubes are generated with an energy spectrum that extends from the beam energy downward and, with high-energy e-beams, the spectrum can go much higher than the energy of decay gamma rays. Accelerators that produce e-beams and x-rays come in many shapes and sizes; all are more complicated and expensive than x-ray tubes, which operate at lower energies than accelerators. At present, there is a "gray zone" in the energy range from 0.5 MeV to 1 MeV for which it is difficult to build x-ray tubes and for which accelerators are not typically constructed, although accelerators could be built to cover this range if there were a market for them. Neutron generators rely on accelerated particles to drive nuclear fusion reactions (typically deuterium-tritium, or D-T, reactions) that release neutrons. These neutrons are highly energetic (14.1 MeV for D-T neutrons) compared to neutrons from radionuclide sources (6 MeV average for americium-beryllium; 2.5 MeV average for californium-252). The intensity of the neutron flux (neutrons emitted per second per unit area) from current neutron generators is lower than is desired, but it has been improving.

Radiation detectors are critical components for replacement of radiation sources for some applications of radiation sources, particularly radiography and well logging. This is because more efficient detectors could enable radiographers and well loggers to use lower activity radionuclide radiation sources or substitute machine sources that are unable to generate the flux achieved with high-activity radionuclide radiation sources. Detectors, too, are improving, but not primarily in response to desires to reduce the use of high-activity radiation sources.

ACCELERATOR TECHNOLOGIES

An atomic particle accelerator—a category of high-energy equipment, which includes linear and circular machines such as betatrons, synchrotrons, cyclotrons, synchrocyclotrons, microtrons, and radiofrequency linear electron accelerators (rf linacs)—is one of the most important tools that modern science possesses. The very high velocity particles produced by these accelerators can be used to break apart atoms, allowing scientists to probe the fundamental principles of matter and energy, and it was primarily for this purpose that these machines were developed. The industrial and medical applications of the accelerators are

[1] This report is concerned mostly with nuclear decay producing alpha, beta, and gamma radiation and neutron emissions from spontaneous fission.

classic examples of basic research translated to significant, unforeseen development of commercial markets.[2]

This chapter describes types of electron accelerators and x-ray and neutron generation techniques, as well as detector technologies that might replace or reduce radionuclide sources in certain applications (see Chapter 1). The accelerators that might be used in these applications are summarized in Table 4-1. The applications and specific replacement technologies are described in greater detail in Chapters 5 through 9.

As illustrated in Table 4-1, there are a many choices of accelerator technologies and a range of possible configurations. The majority of the replacement technologies are based on electron accelerators, which can be configured to deliver either electron beams (e-beams) or x-rays. In many applications, x-rays can be advantageous because of their much greater penetration depths. However, because of the poor energy conversion efficiency in generating x-rays, the required beam powers for electron and x-ray beams differ by more than an order of magnitude for the same radiation dose rate.

It should be noted that there are many other industrial applications that have been made possible because of the properties of electron accelerators or where the clear advantages of the accelerators have already made them, rather than radionuclide radiation sources, the technology of choice. These include applications in material processing such as cross-linking of polymers or curing of composites where the high energy density in an accelerator beam is required (see, e.g., Masefield, 2004). Another application is cargo inspection where high energy density and high x-ray energy are desired to penetrate dense objects.

TABLE 4-1 Summary of Radionuclide Source Applications and Possible Accelerator Replacements

Application	Dose (Gy)	Accelerator Type	Radiation	Energy (MeV)	Power (kW)
Radiotherapy	Few	rf linac	E-beam or x-ray	≈2–30	≈1
Self-contained irradiators	≈1–25	X-ray tube	X ray	≈0.1– 0.4	≈1
Panoramic irradiators	≈100–25,000	dc linac; rf linac; Rhodotron	E-beam or x ray	≈5–10	≈10–1,000
Oil well logging		Electrostatic D-D or D-T	Neutrons	Accelerator produces ≈ 0.1 deuteron or triton for D-T, 2 deuteron for D-D; 2.45 or 14.1 neutron output	≈0.001
Radiography	<1	X-ray tube; Betatron; rf linac	X-ray	≈0.1–20	≈0.001–1

NOTES: Gy = the dose unit, gray; rf = radio frequency; dc = direct current; D-D = deuterium-deuterium fusion reaction; D-T = deuterium-tritium fusion reaction; MeV = mega-electron volt; kW = kilowatt.
SOURCE: Table provided by the committee.

[2] For a general description of the commercial uses of industrial accelerators, see Berejka (1995).

Electron Accelerators and X-Ray Sources

Electron accelerators generate high-energy beams that can be used to directly irradiate an object or to generate x-rays as illustrated in Figure 4-1. Electron accelerators have operated at energies as high as 100 GeV. However, for most applications considered here, the useful energy range is from a few hundred kilo-electron volts (keV) to a few tens of mega-electron volts (MeV). There are many commercial manufacturers of electron accelerators for a wide variety of applications including radiography, materials processing and diagnostics, and medical diagnostics and treatment.

An electron accelerator consists of three parts: an electron source or gun, the accelerator, and a target or scanning/focusing system. Figure 4-1 is a diagram of a medical electron accelerator with an x-ray conversion target. Although Figure 4-1 provides a general layout of linear accelerator (linac) components, there are significant variations from one commercial machine to another, depending on final electron-beam kinetic energy and on the particular design used by the manufacturer.

The length of the accelerating waveguide depends on the final electron kinetic energy, and ranges from about 30 cm at 4 MeV to about 150 cm at 25 MeV. In many applications such as materials processing, food irradiation, and some cancer therapies, the electron beam is sent directly into the object. The beam characteristics at the object are adjusted with bending or focusing fields which, depending on the beam energy, can be either electrostatic or magnetic. In other cases where greater penetration into the object is desired, the electron beam is directed onto a dense target to generate x-rays and the object is irradiated with the x-ray beam. The energy deposition as a function of depth in water is shown in Figure 4-2 for different energy electrons, x-rays, neutrons, and heavy charged particles.

In 2002, it was estimated that there were over 17,000 accelerators in use in industrial or medical applications around the world (Maciszewski and Scharf, 2004). In this chapter, the committee does not discuss the majority of these applications, but concentrates *only* on those applications where the use of radionuclide sources is common.

X-rays are generated through a process known as bremsstrahlung in which the electrons scatter inelastically off heavy atoms in a target; targets typically are made of tungsten or tantalum. This process is relatively inefficient because much of the electron beam energy is deposited in the target itself, although the process becomes more efficient at higher electron energies. For example, the optimized conversion efficiency of a 1-MeV electron beam into x-rays is only about 1–2 percent, whereas this increases to about 8 percent at 5 MeV and to about 12 percent at 7.5 MeV. The x-rays generated through bremsstrahlung have a continuous energy distribution extending down from the electron beam kinetic energy with an average energy that is much lower, typically 20–30 percent of the beam energy; an example of a bremsstrahlung spectrum produced by a 5-MeV electron beam striking a tungsten x-ray target is shown in Figure 4-3. The shape of the bremsstrahlung spectrum as well as the conversion efficiency depend on the kinetic energy of the electron beam striking the target and the thickness and atomic number of the target. The x-ray beam of Figure 4-3 would be referred to as a 5-MV x-ray beam where MV stands for "megavoltage" and the term 5 MV implies that we are dealing with an x-ray bremsstrahlung spectrum that contains photons with energies from 0 to 5 MeV and is produced in the x-ray target by monoenergetic electrons with a kinetic energy of 5 MeV.[3]

[3] By convention, x-ray beams in the megavolt range are abbreviated as MV, but those in the kilovolt range are abbreviated as kVp

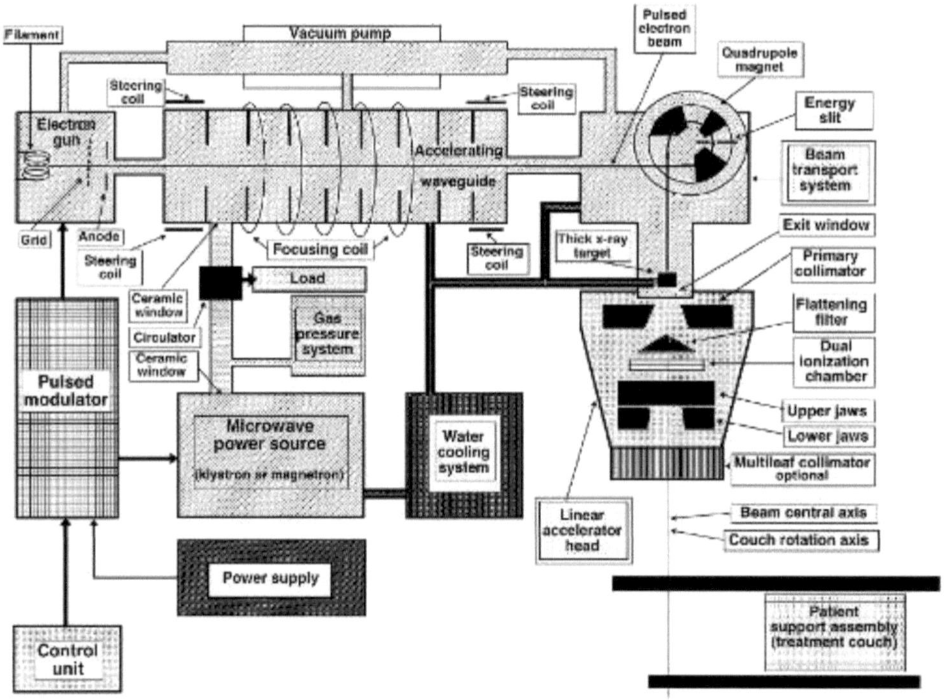

FIGURE 4-1 Schematic diagram of a typical medical electron accelerator with x-ray conversion target. SOURCE: Image provided by committee.

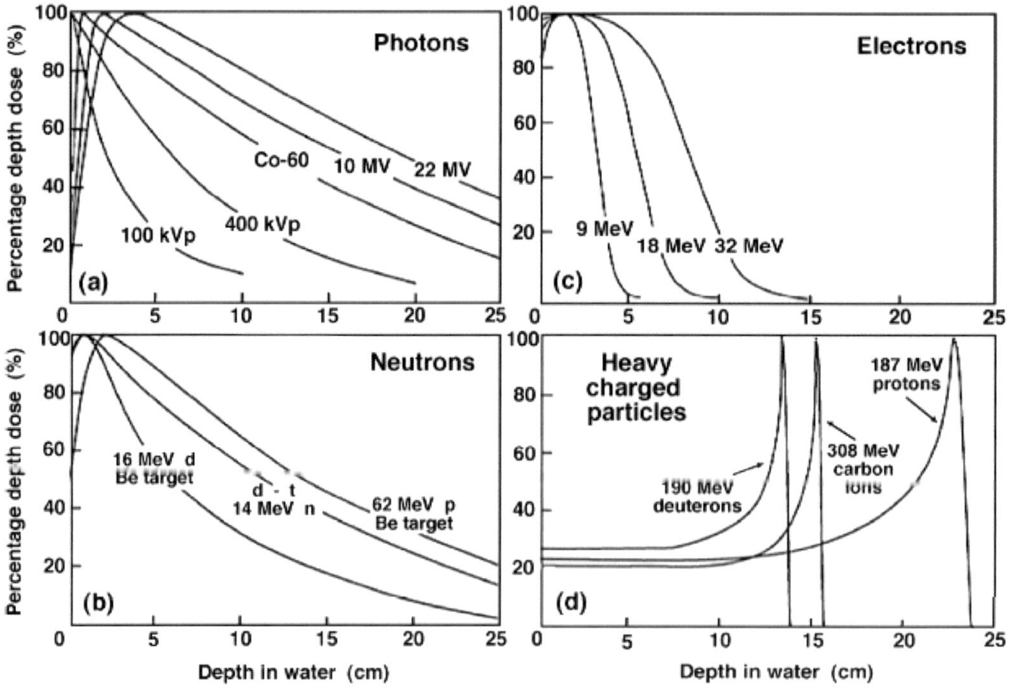

FIGURE 4-2 Absorbed dose plotted as a function of depth in water for ionizing radiation beams of various types and energies. Parts (a) and (b) are for indirectly ionizing radiation: in (a) for photon beams in the range from 100 kVp to 22 MV and in (b) for neutron beams. Parts (c) and (d) are for directly ionizing radiation: in (c) for megavoltage electron beams in the range from 9 to 32 MeV and in (d) for heavy charged particle beams (187 MeV protons, 190 MeV deuterons, and 308 MeV carbon ions). SOURCE: With kind permission of Springer Science+Business Media.

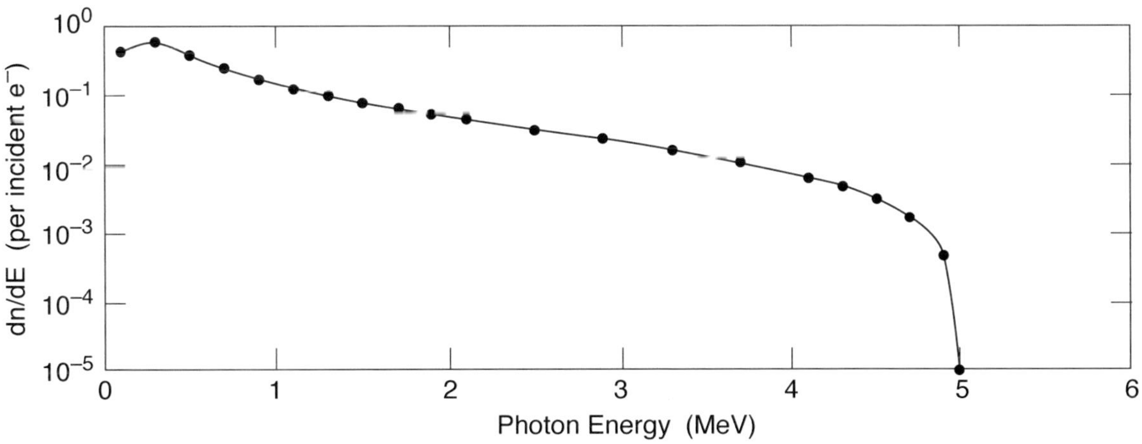

FIGURE 4-3 Bremsstrahlung x-ray (photon) energy spectrum from a 5-MeV electron beam. The ordinate (vertical axis), *dn/dE,* shows the number of x-rays within a certain energy range per incident 5-MeV electron. SOURCE: Image provided by the committee.

The accelerator-based sources described here are grouped into three categories by their acceleration method: high-voltage dc accelerators, rf microwave accelerators, and induction accelerators; x-ray tubes are a special type of dc accelerator. Beam energies between 100 keV and about 30 MeV are potentially useful for most applications that are considered in this report. Typical medical accelerators operate with beam powers of roughly 1 kW; low-power accelerators for radiography may only have beam powers of a few watts; and high-power accelerators for irradiation operate with average beam powers of tens to hundreds of kilowatts. In high-power applications, such as those needed in large irradiators, the beam energy is limited by neutron production and activation of the accelerator and the target materials. The threshold for neutron production is between 8 and 13 MeV in most materials. In the case of food irradiation, the U.S. Food and Drug Administration limits the electron-beam energy to 10 MeV for electron irradiation and 7.5 MeV for x-ray irradiation (Meissner et al., 2000). Radiation hazards and radioactive waste from accelerators are discussed later in this chapter.

The energy and beam power requirements for the accelerator system depend on the size and type of radionuclide source that it is replacing. To set the scale, the radiation dose from a radiotherapy source with 370 TBq (10,000 Ci) of cobalt-60 is roughly equivalent to the x-ray dose from a 10-MeV linac with 100 watts of beam power. A portable radiography source with 11 TBq (300 Ci) of iridium-192 could be replaced using the x rays generated by a 1-MeV electron beam with a beam power of a few watts. On a larger scale, the radiation dose in a panoramic irradiator using 110,000 TBq (3 MCi) of cobalt-60 would be roughly equivalent to that from a 20-kW, 10-MeV electron beam or an x-ray beam generated from a 300-kW, 7.5-MeV electron beam.

High-Voltage dc Electron Accelerators

Direct current (dc) accelerators use a dc voltage to accelerate an essentially continuous wave (cw) beam. The beam is accelerated with an electric field generated with a series of electrodes. To prevent breakdown (direct electrical discharge that short-circuits the voltage gap), the fields are typically limited to much less than 1 MV/m and the energy of the electron beam is usually limited to a few MeV although dc devices have been operated at energies as high as a few tens of MeV. Because of the breakdown limitations, these devices tend to be larger than rf accelerators. Examples of dc accelerators include x-ray tubes as well as devices for materials analysis and processing, semiconductor processing and development, and pharmaceutical research. These devices can have an output power as high as a few hundred kilowatts, and high-efficiency solid-state switching techniques may further improve the power output to a few megawatts.

High-voltage dc generators are based on either mechanically transporting charges, as is done in a Van de Graaff or Pelletron accelerator (referred to as electrostatic accelerators), or cascaded rectifier circuits (referred to as electrodynamic accelerators; Norton and Klody, 1997). Both types of dc accelerators are produced by many manufacturers including IBA, Nissin High Voltage, Pelletron, and Vivirad. These accelerators operate at voltages between 100 keV and 10 MeV and are frequently used for electron-beam irradiation in materials processing, for example, to improve the physical properties of plastics, cables, and wires, or for materials analysis; similar sources are also used at a much lower energy for ion implantation in semiconductor development. In general, the electrodynamic accelerators produce greater beam powers and are used for materials processing applications.

Low-energy dc accelerators (approximately 100 to 500 keV) have been constructed with a few megawatts of beam power. These high-power[4] accelerators are primarily used for materials processing applications. Examples of higher voltage accelerators include the Dynamitron from IBA (Cleland, 1959; see also Figure 4-4) and the ICT from Vivirad, both of which operate at up to 5 MeV with about 200–300 kW of beam power.

The dc accelerators are relatively straightforward to control, and the beam voltage and current can be varied over large ranges. For example, a Dynamitron is typically designed to be operated over a range from 1 to 3 MeV at constant beam current, and a much wider range (e.g., 300 keV to 3 MeV) if the beam current is lowered. Such large variation can be more difficult to attain in rf accelerators.

The improvement and high power capability of electronics and semiconductor switches, in particular the insulated gate bipolar transistor (IGBT) and the integrated gate commutated thyristor (IGCT), will likely lead to improved performance in the dc accelerators. At present, the typical operating efficiencies are roughly 70 to 80 percent. New multimegawatt power supplies (see, e.g., Bradley et al., 1999; Cassel et al., 1997; and Diversified Technologies, Inc., 2007) with voltages of about 100 kV are being developed for other applications that have efficiencies in the high 90 percents; it is expected that these concepts will also lead to improvements in the higher voltage dc accelerators as well.

X-ray Tubes

X-ray tubes are low-energy dc accelerators in which the electron gun (cathode), accelerator, and target are contained within a single vacuum enclosure. Tube voltages can range from a few kilovolts (kV) up to about 500 kV and are used in many applications ranging from medical devices to irradiators to industrial materials processing; x-ray tubes are manufactured by a wide range of producers in the United States and abroad.

The simplest x-ray tubes are evacuated closed glass tubes with a tungsten filament (cathode) that emits electrons and a tungsten target that generates x-rays. The lifetime of a tube can be limited by the tungsten filament, tungsten plating that forms on the glass and leads to arcing, or the deterioration of the vacuum within the tube. Typical lifetimes of closed glass tubes are about 1000 to 5000 hours of operation. Closed metal and ceramic tubes can achieve significantly longer lifetimes; at least one company manufactures a closed x-ray tube with a

---

[4] Note that there is a difference between energy and power. Energy refers to the energy of the individual particles. Power is the amount of energy that flows through the beam in a given amount of time. An accelerator that produces an intense beam of low-energy particles can be a high-power machine.

dispenser cathode[5] rather than a tungsten filament and the tube has a rated lifetime of 20,000 hours of operation. Open or dismountable tubes have a separate vacuum pump which allows the tube to be opened and the cathode and anode to be replaced as necessary. With the appropriate preventive maintenance, dismountable tubes can be operated for years without failure.

In high-power applications, the anode must be constructed of high-temperature materials because, as noted earlier, the x-ray generation process is inefficient and most of the electron beam energy is deposited into the target. Typical fixed anode tubes are limited to powers of a few kilowatts. A further refinement is a rotating anode tube, where the target is rotated to spread the heat deposition around a larger area. Such tubes can operate at power levels many times higher than the fixed anode tubes.

Much of the development of robust x-ray tubes is being driven by the medical imaging and nondestructive testing (NDT) requirements. In both of these applications, high power and long tube lifetime are important. Additional refinements such as cold (field emission) cathodes have increased the energy and power available in compact and miniature x-ray tubes (see Reyes-Mena et al., 2005; Xintek Inc., 2004–2006), which make them an attractive alternative in situations where small radiation sources are required. Low-energy (40 to 50 kVp) miniature x-ray tubes with hot cathodes have also been developed as brachytherapy sources.

FIGURE 4-4 A Dynamitron showing the accelerator column. SOURCE: Image courtesy of ION Beam Laboratory, State University of New York at Albany.

---

[5] A dispenser cathode is a tungsten matrix doped or impregnated with a material, such as barium or ruthenium oxide, that lowers the work function or energy required to liberate an electron from the surface of the cathode. Such cathodes can generate the same current at lower temperatures than standard tungsten filaments, which extends the operating lifetime.

RF Electron Accelerators

Radiofrequency accelerators accelerate beams with rf cavities that typically operate at low microwave frequencies (on the order of 1 GHz), although there are accelerators based on lasers (about 100 THz) and relatively low-frequency rf accelerators, which operate in the 100-MHz regime. Usually the rf cavities are limited to accelerating fields of a few tens of megavolts per meter (MV/m). Laser-based accelerators have achieved fields as high as 100 GeV/m, but these are far from being commercial devices. Unlike the dc accelerators, rf accelerators do not have to prevent discharge across the full acceleration voltage along their length, which allows rf accelerators to operate over a large energy range and be relatively compact. There are many different variants of rf accelerators; examples include linear accelerators for medical applications (on the order of 1–10 MeV), synchrotron storage rings for synchrotron radiation generation (on the order of 1 GeV), and high-energy synchrotrons or linacs for high-energy physics (on the order of 1 TeV). Radiofrequency accelerators may be single-pass linacs or multiple-pass accelerators such as microtrons or the Rhodotron, or circular accelerators such as cyclotrons or synchrotrons. These are described below, except for circular rf accelerators, which are used in proton and carbon ion radiotherapy but are not typically used in the energy range of interest as potential radionuclide radiation source replacements.

Radiofrequency Linacs

Radiofrequency linacs were developed in the 1940s and are used for many applications ranging from the generation of x-rays in a hospital environment to injectors into higher energy synchrotrons at particle physics laboratories. The first rf linacs were built to operate around 3 GHz (similar to the frequency used in household microwave ovens and in some cordless phones), and today most commercial linacs operate at frequencies between 1 and 15 GHz with rf wavelengths that are between 30 and 2 cm, respectively.[6]

A large number of companies manufacture rf linacs for a variety of applications. The most common use of rf linacs is for medical radiotherapy. A number of companies manufacture such accelerators, including Elekta, Mitsubishi, Siemens, and Varian. The two other major applications that are considered in this report are non-destructive testing (radiography) and irradiation/sterilization. A number of companies, both large and small, produce linacs for these applications; a few examples include AS&E, Hitachi, L3 Communications, Linac Technologies, L&W Research, Tsinghua Tongfang Nuctech, and Varian. A few commercial rf linacs are illustrated in Figure 4-5.

An rf linac is constructed from four main elements: (1) a high-voltage power supply (modulator), (2) an rf power source, (3) a microwave cavity, and (4) a charged particle source as illustrated in Figure 4-1. The microwave cavity is the "heart" of the accelerator. It is constructed from a series of cavities with an aperture along the axis for the beam. The size of the cavities is selected on the basis of wavelength (and therefore the rf frequency) of the linac, and is independent of the overall size of the accelerator. For example, the individual 3-GHz cavities in the Stanford Linear Accelerator Center (SLAC) linac are roughly 10 cm (4 in.) in diameter and about 3.8 cm (1.5-in.) thick with a 2.5 cm (1 in.) diameter hole passing through the center through which the electron beam and rf power pass. In SLAC, roughly 90 individual cavities are bonded together for each of its many 3-meter-long accelerator structures, whereas a lower energy linac might have just one accelerator structure with a small number of cavities. Because

[6] The ILU-series of linacs operate with rf frequencies around 100 MHz (Auslender, 2005).

the cavity dimensions scale with the rf wavelength, a similar structure operating at 12 GHz would be roughly one-quarter the size.

Typical acceleration gradients scale with frequency due to electrical breakdown and power limits. Normal conducting 1-GHz linacs typically have maximum gradients that are about 10 MV/m, while at 12 GHz, linacs have operated with gradients greater than 70 MV/m (Adolphsen et al., 2005). However, the rf power required to achieve these high gradients is quite high. High gradients are desirable because they make the linac shorter for a given particle energy and they give less travel time for the beam to spread as a result of the particles electrically repelling each other.

In cases where high radiation doses are desired, such as for irradiation facilities, the accelerator efficiency is very important. The overall electrical efficiency from power source to emitted beam power of early rf linacs was only about 20 to 30 percent, compared with efficiencies of 60 to 80 percent for the dc accelerators. The efficiency can be improved by using more efficient rf power sources and maximizing the efficiency of the accelerator cavity by choosing the beam current so that the beam-induced voltage is comparable to the unloaded voltage—the optimization depends on the detailed cavity design. For normal conducting cavities, this implies beam currents that are comparable to 1 ampere and the resulting rf-to-beam transfer efficiencies can be as high as about 70–80 percent; some examples are listed in Haimson (1975) and Miller et al. (2003). Superconducting cavities can be optimized so that essentially all of the rf energy is transferred to the beam; however, the technology is still relatively novel and there are no commercial superconducting linacs presently operating.

The simplest and most common particle source for an rf linac is a thermionic gun where a relatively low dc voltage of about 20 kV is used to accelerate electrons from a heated cathode. This type of gun has the disadvantage that the beam is not appropriately bunched for the rf linac, which results in a broad energy spectrum, particle losses, and a relatively large beam diameter. Variations on the thermionic gun include higher voltage operation (up to 500 kV), the use of high-frequency choppers and subharmonic bunchers to improve the bunching and capture of the electron beam, and voltage grids to clearly define the length of the pulse (from 1 nanosecond to a few microseconds). In most cases, these additional complications are not useful for the applications considered here.

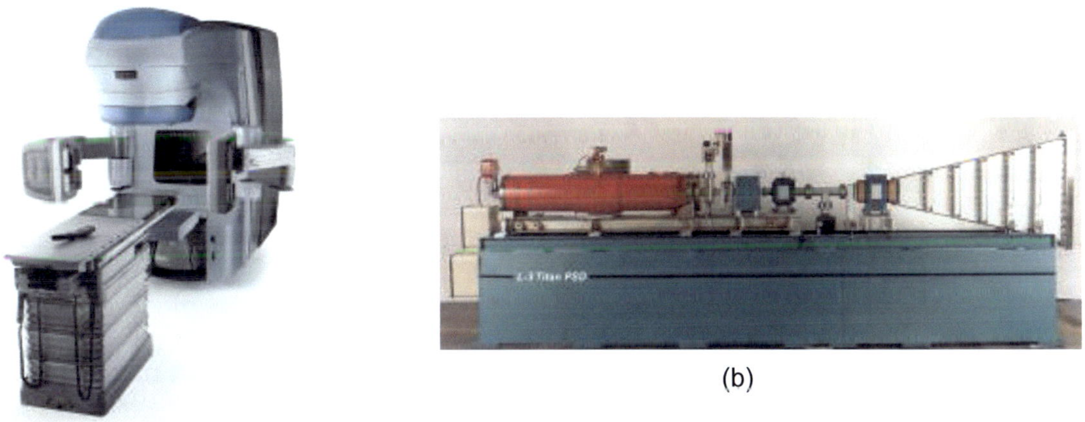

(a) (b)

FIGURE 4-5 Examples of commercial rf linacs. SOURCE: (a) Courtesy of Varian Medical Systems, Palo Alto, CA; (b) courtesy of L3 Communications.

Another type of injector is the rf gun where high-voltage microwave fields accelerate the beam right from the cathode. For low-energy applications of a few MeV, the gun can be integrated onto the accelerator structure to directly deliver the desired beam. Because the beam is rapidly accelerated, many of the effects that limit the dc gun performance due to the self-repulsion of the electrons are mitigated. Radiofrequency guns have been developed using lasers and photocathodes to generate short-pulse electron beams (a few picoseconds) which are matched to the rf wavelength of the accelerator (Sheffield et al., 1988). Photocathode rf guns tend to have extremely high quality beams but require a costly laser system. Radiofrequency laser cathodes are not commercially available. Thermionic rf guns that eliminate the laser system also have been developed (Beczek et al., 2001; Tanabe et al., 1989; Westenskow and Madey, 1984). Thermionic rf guns are becoming more common as the technology becomes more mature and will likely find use for the applications considered in this report.

The rf power source that generates the microwaves for the linac is usually either a magnetron oscillator or a klystron amplifier. In both cases, the size of the rf source and the power output capability are roughly proportional to the rf wavelength. Magnetrons were developed during the 1940s and are used in everything from microwave ovens to sophisticated radar systems. Magnetrons are relatively simple and compact but have limited output power and limited control over the rf frequency and phase. Continuous-wave devices can have an output power as high as about 100 kW at 1 GHz with efficiencies of about 75–85 percent while pulsed devices can operate at about 60–75 percent efficiencies. To be an effective power source for an rf linac, the magnetrons usually need a feedback system to stabilize the rf output. Klystrons are high-power amplifiers that tend to have higher power capabilities than the magnetrons but are also larger, heavier, and have lower efficiencies. Continuous-wave klystrons can have output powers of around 1 MW at 1 GHz with efficiencies of 60–70 percent while pulsed devices with efficiencies of about 50 percent have peak powers that range from about 100 megawatts with microsecond pulses to about 10 megawatts with microsecond pulses. New high-efficiency klystrons that are being developed in laboratories and industry use configurations based on planar beams or multiple round beams to reduce the repulsive forces within the beam; efficiencies of about 65 percent are expected for the pulsed klystrons (sheet-beam klystrons; multibeam klystrons; for examples, see Lenci et al., 2004). The rated lifetimes of typical high-power magnetrons or klystrons is usually 5,000 to 20,000 hours. Over the past two decades, the lifetime of typical magnetrons has been extended by more than a factor of three. It is expected that modern design will further improve both the efficiency and reliability of these rf power sources (Vlieks et al., 1998).

Finally, the modulator or high-voltage power supply converts the incoming alternating current (ac) voltage into the high-voltage dc power needed for the rf power sources. As noted earlier, the improved high power capability of semiconductor switches, in particular the IGBT and the IGCT, has led to improved power handling, efficiency, and reliability in both modulators and high voltage power supplies. These systems operate with minimal losses and have expected availability of more than 50,000 operating hours, which is many times greater than that achieved with conventional technology.

Like x-ray tubes, the reliability of rf linacs has been driven by the medical and industrial applications. Modern medical linacs have useful lifetimes of 10 to 15 years, after which the computer controls and therapy planning software generally are considered to be obsolete and unable to be upgraded, although the linac may be fully operational. The waveguides and rf cavities tend to be robust against failure and, as described above, the lifetimes, as well as the mean times to repair, of the consumable components, the electron gun, rf power source, and voltage supply have been greatly improved over the past few decades.

Multipass rf Accelerators

There are two types of multipass rf accelerators that are most useful in the energy range of interest for this study: microtrons (Veksler, 1944) and the Rhodotron (Jongen et al., 1993; Pottier, 1989). A microtron uses a standard accelerator cavity to accelerate the beam. The beam is recirculated with a set of bending magnets in either a circular or racetrack configuration (see Figure 4-6). Typical designs use tens of passes through the same accelerator cavity to accelerate the beam and, depending on the energy gain per turn, the output beam energies can range from a few MeV to 1 GeV. Because the accelerator cavities are used for multiple passes, microtrons have the potential for being less expensive than linacs, although the bending magnets that recirculate the beam can be expensive. In addition, because the microtron can operate in a cw mode, the output powers can be more than 100 kW. There are a number of commercial manufactures of microtrons, including Scanditronix Medical and Sumitomo Heavy Industries.

The Rhodotron is also a recirculating accelerator where a very low frequency field is used to accelerate the beam radially through a cylindrical cavity with 5 to 10 passes. The low-frequency operation (about 100 MHz) has low losses and the rf power can be generated with an efficient high-power tetrode (a four-element electron tube). Commercial Rhodotrons are produced by IBA Industrial and operate at energies up to 10 MeV with beam powers as high as 700 kW (IBA Industrial); one such device is shown in Figure 4-7. The Rhodotron operates with roughly 50 percent efficiency at peak output power.

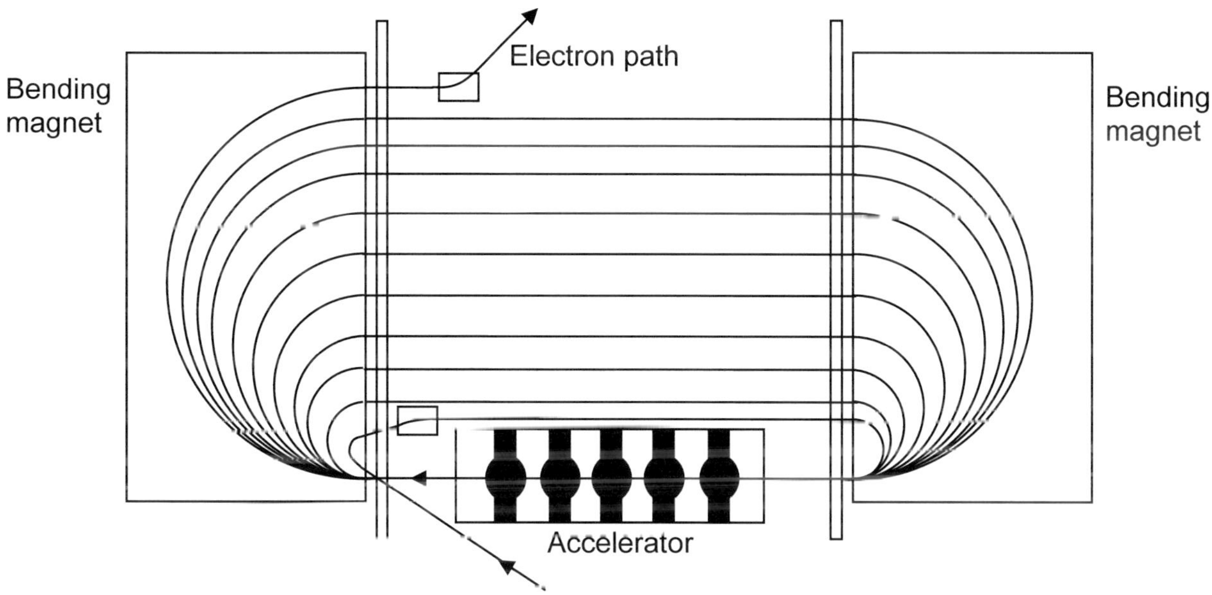

FIGURE 4-6 Diagram of a microtron. SOURCE: Image provided by the committee.

FIGURE 4-7 This 200-kW Rhodotron is used to irradiate U.S. mail at a facility in New Jersey. SOURCE: Image courtesy of IBA-RDI.

Induction Electron Accelerators

Induction accelerators operate using magnetic induction where the accelerating voltage is generated by the changing magnetic field. The two types of induction accelerators are betatrons (Kerst, 1940) and induction linacs (Christofilos et al., 1964). The beam current in a betatron tends to be limited by space charge effects (self-repulsion of the electrons) and they operate as pulsed devices; these are relatively low power accelerators. Betatrons were used for radiotherapy (see, e.g., Kapetanakos et al., 1993); however, these have largely been replaced with rf linacs, which provide greater beam power in a more flexible package. Presently, betatrons are primarily manufactured as portable devices for nondestructive testing (see, e.g., Inspecta, 2007; JME Ltd, 2007; Kaplin et al., 2002) and have energies of a few MeV with a radiation output of a few roentgen per minute. Figure 4-8 is an illustration of such a device.

In contrast, induction linacs can accelerate high-current beams. They are inherently low-impedance devices and thus are used to produce low-energy (a few MeV), high-current (kA) pulses at a relatively low repetition rate. They are primarily used for high-intensity flash radiography, inertially confined fusion drivers, and directed energy weapons. Examples include the AIRIX accelerator at Centre d'Etudes Scientifiques et Techniques d'Aquitane (Eyharts et al.,1995) and the DARHT accelerator at Los Alamos National Laboratory (Burns et al., 1996). Because the high peak power is not necessary when considering radionuclide source replacement and induction linacs are relatively expensive for a given average power, they are not considered as potential radiation source replacements in this report, although additional induction linac research and development is mentioned below.

FIGURE 4-8 The JME Betatron Data Pack for industrial radiography. SOURCE: Image Courtesy of JME, LTD.

New Developments in Electron Accelerators

A number of cutting-edge research demonstrations that could improve accelerator capabilities have been carried out or are underway. These are described below. Although none of these is close to commercial deployment, they illustrate the potential for higher efficiency, higher power, more compact, and specialized radiation generators.

High-gradient superconducting rf accelerators have been developed over the past few decades and can provide a very efficient method of accelerating an electron beam. A number of accelerators are using superconducting rf cavities, including the recirculating linac used for nuclear physics experiments at the Jefferson Laboratory in Virginia and the FLASH UV laser facility at the DESY laboratory in Hamburg, Germany. These are research and development programs that indicate possible future directions. Recently, pulsed rf cavities, operating at about 1 GHz and 1 m long, have demonstrated gradients in excess of 35 MV/m, and gradients in excess of 15 MV/m have been maintained in continuous wave operation. Such cavities are being industrially produced by companies such as Advanced Energy Systems in the United States and ACCEL in Europe, which was recently purchased by Varian Medical Systems. However, turnkey superconducting linacs are not yet available.

High-power electron guns using superconducting rf cavities are being developed for various purposes. These guns are designed to produce very high quality continuous wave beams with beam energies of a few MeV and beam powers of 100 to 1,000 kW. For example, one gun presently being fabricated by a collaboration between Jefferson Laboratory and Advanced Energy Systems is designed to produce a 7-MeV beam with a current of 100 milliamperes; this device is presently being commissioned at lower power levels. Similar devices are being designed and constructed at many laboratories around the world as injectors for energy recovery linacs. Such accelerators may provide efficient high-power beams for irradiation and sterilization facilities in the future.

The primary disadvantage of the superconducting rf technology is that additional cryogenic cooling is required, which increases the capital cost of a new facility. For best performance, the superconducting rf cavities typically are operated at about 2 K (–271°C). For a large installation, the incremental cost and additional complexity in the cryosystem is relatively small, but it may not be cost-effective for a smaller stand-alone facility. Additional analysis is needed to understand these cost trade-offs.

Other accelerator developments include:

1. Work on more cost-efficient high-gradient induction linacs. The advanced solid-state switches are leading to improved efficiency and lower cost per module. There is active research and development on the use of dielectric-loaded structures which could permit about 10 times higher gradients, approaching 20 MeV/m (Sampayan et al., 2005). This would make the high-power characteristics of the induction technology much more attractive.
2. Development of compact portable mega-electron volt accelerators that could replace hand-portable iridium-192 sources for radiography (Breidenback, M., Stanford Linear Accelerator Center, private communication with Raubenheimer, T., 2007; Yamamoto et al., 2006).
3. Demonstration of high-gradient laser accelerators that have the potential to make a very compact source of medium-energy (about 20–100 MeV) electrons (Faure et al., 2006).
4. Development of narrow-band x-ray sources based on Compton backscattering or crystal diffraction. Some of these sources are available commercially (Vlieks et al., 2006; Dobashi et al., 2005; Mondelaers et al., 2000; Jolie et al., 1998).

ACCELERATOR-DRIVEN NEUTRON SOURCES

Particle accelerators can also be used to generate neutrons. Setting aside spallation sources, which require much higher energies (GeV) and therefore large facilities, accelerator-driven neutron sources direct a beam of deuterium nuclei at a target loaded with deuterium or tritium,[7] causing fusion reactions. The deuterium-deuterium reaction (D-D reaction), the deuterium-tritium (D-T reaction), and a third reaction that is not typically used, the tritium-tritium (T-T reaction), are illustrated below in Equation 4-1. The reactions produce helium nuclei and neutrons:

$$\begin{aligned} D + D &\rightarrow {}^{3}He + n, \\ D + T &\rightarrow {}^{4}He + n, \\ T + T &\rightarrow {}^{4}He + 2n \qquad (4\text{-}1). \end{aligned}$$

The D-D reaction generates monoenergetic neutrons at 2.45 MeV, but has a competing reaction that generates no neutrons. The neutrons produced by the D-T reaction are emitted monoenergetically at 14.1 MeV. Laboratory-scale accelerator-driven fusion sources based on the D-T reaction, such as the Rotating Target Neutron Source, have been in operation for several decades (see, e.g., Booth, 1967) and generate moderate neutron flux (e.g., 5×10^{11} neutrons per square centimeter per second). Kaman Nuclear, formerly in Colorado Springs, produced neutron generators from at least 1963 with advertised flux ratings ranging from 10^7 to 10^{11} neutrons per square centimeter per second. The T-T reaction generates neutrons with energies between about 1 MeV and 10 MeV, and averaging about 5 MeV. Figure 4-9 shows the cross section, which scales directly with the reaction rate, for several fusion reactions. The D-T cross section peaks below 100 keV in center of mass coordinates, which translates into around

[7] Deuterium and tritium are isotopes of hydrogen: hydrogen-2 and hydrogen-3, respectively.

110 keV for a deuteron from an accelerator striking a stationary tritiated target. The peak D-D cross section is more than an order of magnitude lower and peaks closer to 1 MeV in center of mass coordinates.

The first commercial pulsed neutron well logging tool was introduced in 1963 by Dresser Atlas, but its capabilities were very limited because the neutron flux was quite low. Within the past 10 to 15 years, higher flux compact neutron generators have become technically feasible. Although several companies manufacture compact neutron generators for use in well logging and other applications, the units remain quite costly. They are manufactured by the All-Russia Research Institute of Automatics (VNIIA); Baker Hughes, Inc.; China Petroleum Technology and Development Corporation; Eads Sodem; Halliburton Company; Schlumberger, Ltd.; and Thermo Electron Corporation. Figure 4-10 shows a schematic of the unit designed by Sandia National Laboratories and manufactured by Thermo Electron Corporation.

Typical neutron tubes generate 10^6 neutrons/pulse and pulse 100 times per second to yield 10^8 neutrons/s. Higher neutron output rates (10^{11} neutrons/s) can be achieved in some larger units. Several neutron generators with neutron outputs ranging from 10^7 to 10^{14} neutrons/s have been developed (Lawrence Berkeley National Laboratory). One of these approaches uses a low-temperature tritium plasma formed inside a tubular deuterium-loaded target and the ions are accelerated toward the walls where the reactions take place.

Another promising idea that has been suggested is essentially the inverse of this approach, using plasma immersion ion implantation techniques (I. Brown, Lawrence Berkeley National Laboratory, personal communication with M. Lowenthal, May 8, 2007). In this idea, a low-temperature tritium plasma would be formed around a deuterium-loaded target. The target would then be pulsed to a voltage of −100 kV, and the ions would strike the target. Like the preceding approach, this technique is not limited by the intensity achievable in an ion beam and could lead to higher intensity neutron sources with microsecond pulse widths. To the committee's knowledge, this approach has not been examined in detail.

Reaction rates, and therefore the neutron output rates, could in principle also be raised by developing targets that hold higher concentrations of hydrogen isotopes.

RADIATION HAZARD AND RADIOACTIVE WASTE FROM ACCELERATORS

It must be noted that switching from radionuclide radiation sources to machine sources of radiation does not obviate the need for radiation safety and radioactive waste management. Particle accelerators emit radiation primarily along their beamlines, but also to much lesser extent they emit x-rays in other directions. Where and how much radiation is emitted depends on the design of the accelerator. Operators of accelerators and equipment that uses x-ray tubes are required to undergo radiation safety training. The ability to turn off a radiation generator is an obvious advantage for worker safety.

Particle accelerators also can, and do over time, induce low but measurable concentrations of radioactive material in their bremsstrahlung targets and other objects subjected to extensive irradiation by electrons or photons in the several-MeV range and higher through photonuclear reactions (see NCRP, 2005, 1984, 1977).

Accelerators that operate at energies below 10 MeV pose fairly insignificant radiation hazards when they are not operating. Neutron generators induce radioactivity in the material surrounding the fusion reaction (the target, the device, the housing, and any shielding) through neutron capture and other nuclear reactions. The fusion target or source plenum contains tritium, which is itself radioactive, and the neutron generator becomes contaminated and must be treated as radioactive waste. The radioactive material imposes only a minor waste management burden, as the concentrations of radionuclides are typically very small compared to Category 1 and 2 radionuclide radiation sources.

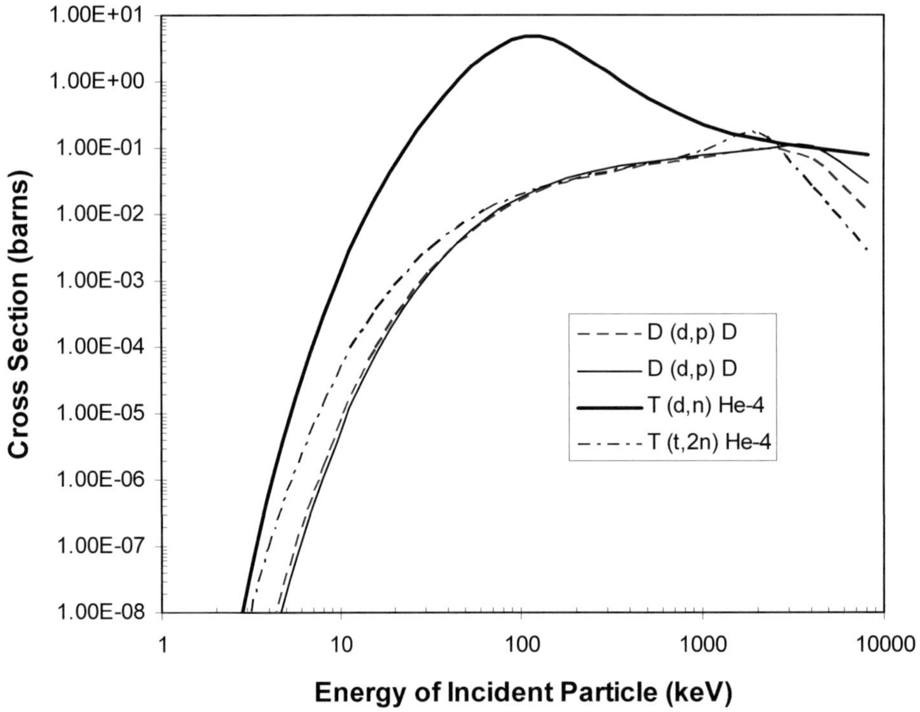

FIGURE 4-9 Fusion reaction cross sections as functions of kinetic energy. SOURCE: Image provided by the committee.

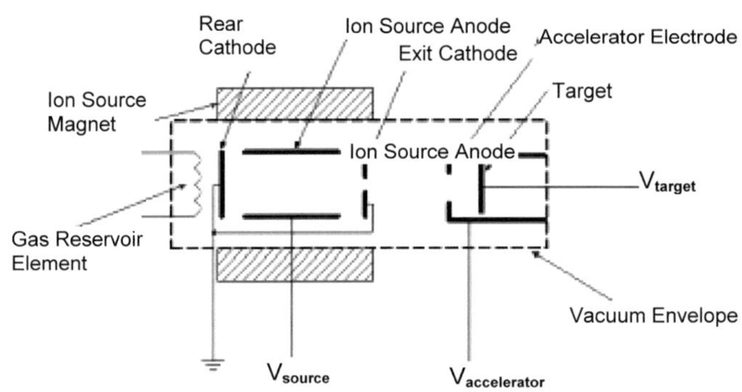

FIGURE 4-10 Schematic of a neutron tube. SOURCE: Image courtesy of Burkhart, B. (2006).

DETECTOR TECHNOLOGIES

Each radiation source application discussed in this report uses the radiation for one of two purposes: to deposit energy within the irradiated material or to gather information about the irradiated material. Radiation sterilization and radiotherapy use the deposited energy to kill bacteria, viruses, or cancer cells; blood irradiators use it to kill white blood cells. This can be accomplished with a radionuclide radiation source or radiation generator working alone. In contrast, radiography and nuclear well logging aim to learn more about the structure and composition of the irradiated material. These latter applications use a radiation source or radiation generator in conjunction with a detector. In gamma radiography the detector is commonly a sheet of film. In x-ray radiography, charge coupled device (CCD, like those used in digital cameras) detectors are increasingly replacing sheets of film. In well logging, the gamma rays triggered by radiation interactions in the rock layers adjacent to the well hole are usually detected with thallium-doped sodium iodide crystal scintillators. How useful an apparatus is depends on the performance of the overall system, not just the source, and so, improving the detector can relax the demands on the source. That is, a more sensitive detector may enable users to accomplish the same tasks with a lower intensity radiation source, including a small radiation generator.

The underlying physical principles of all radiation detectors are the same. Radiation passing through the detector interacts with some material within the detector. Neutral particles, such as gamma rays, x-rays, and neutrons have relatively large ranges in materials, depositing their energy over tens of centimeters of solid material. Charged particles, such as alpha particles and energetic electrons, deposit their energy in microns or millimeters of solid material. In some detection schemes, the interaction is converted into an electrical signal, which can trigger an alarm, be processed for display, or simply be recorded. In some radiation detectors, the electron or the ion directly creates the electrical signal. For example, in a gas ionization chamber or a solid-state detector, a voltage difference is placed across the detector media so that when electrons are liberated and ions (or holes, in the case of a solid-state photoconductor or photodiode) are formed, an electrical current flows. Other radiation detectors use phosphors or scintillators to convert the incoming radiation to visible light or some other relatively long wavelength radiation. The detectors can then use a photomultiplier tube or a CCD to convert light into electrical signals. Detectors have improved over the past several decades, primarily as semiconductors have been applied and improved and as more efficient phosphors and scintillators have been discovered. These improvements enable detectors today to map the radiation spatially, forming an image, and to discriminate energies.

Although it has been documented that x-rays can cause faint visual effects in the human eye, called radiation phosphenes, x-rays and gamma rays are, for practical purposes, invisible. Historically, the first radiation detection devices were photographic plates, modern versions of which, particularly large-format, are still used today in radiography. The sensitivity of the early photographs was improved substantially by the advent of “intensifying screens” placed in front of the emulsion. These screens fluoresced under x-ray irradiation and the light from the screen supplemented the x-ray interaction with the emulsion (Frame, 2004). Faster films replaced the early plates and screens, but the general approach of converting the x-ray “signal” into an intermediate, more readily sensed or recorded form underlies many detection techniques today.

The majority of gamma-ray detectors are scintillator devices; they convert a gamma ray into visible light which in turn is then converted into an electrical signal with a semiconductor detector, as described above. In solid state, the gamma ray is directly absorbed in a p-n junction semiconductor to create an electrical output. (This is analogous to the operation of a solar cell.) The main thrust for the development of gamma-ray detectors until very recently has been for high-energy particle physics experiments where the energies are much higher than those used in radiography and a very wide range in energies must be detected simultaneously. Recently, though, there has been renewed interest in developing detectors for 511-keV photons used in positron emission tomography and 662-keV gamma rays. At the present stage of detector development, the scintillator approach is generally preferred because it enables both the scintillator detector and the light detector to be selected for optimum performance.

The development of detectors for medical computed tomography (CT) x-ray scanners provides an example of how a combination of technological advances has enabled more imaging information to be obtained with the same or reduced source intensity. For CT, a series of x-ray exposures have to be recorded rapidly to obtain sufficient information for the computer reconstruction of a three-dimensional image while minimizing the x-ray exposure of the patient. To minimize exposure, this has necessitated the development of more efficient scintillator materials that can convert each x-ray photon to light while minimizing afterglow in order to capture images as quickly as possible. Different manufacturers have developed different scintillator materials specifically for this purpose, such as Gd_2O_2S:Pr,Ce,F and $(Y,Gd)_2O_3$:Eu,Pr. The common feature of these materials is that they have been designed and doping has been manipulated to optimize the efficient conversion of high-energy x-rays to light that can be most efficiently detected by CCD. Concurrently, there have been dramatic improvements in CCD technology, computer software and compact x-ray sources dedicated to the development of the state-of-the-art CT scans commercially available today.

Recognizing that improved detectors for well logging applications would be beneficial in decreasing the activity of the current radiation sources, several alternative materials, such as bismuth germanate (BGO), have been investigated, but none has yet demonstrated the reliability, reproducibility, and long-term stability that the current alkali-halide (NaI;Tl) exhibits. However, the development of these alternative materials has not been directed specifically for the detection of gamma-ray energies for well logging applications but rather for the detection of gamma rays over a broad energy range.

The detection of neutrons, such as used in well logging, poses particular challenges because their interaction with detectors is so weak. Neutron detectors typically utilize nuclear reactions in helium-3 ($n + {}^{3}He \rightarrow {}^{3}H + {}^{1}H + 0.764$ MeV), boron-10 ($n + {}^{10}B \rightarrow {}^{7}Li + {}^{4}He + 2.31$ MeV), or lithium-6 ($n + {}^{6}Li \rightarrow {}^{3}H + {}^{4}He + 4.78$ MeV). The probability of these reactions occurring with an incoming neutron is high at low neutron energies (fractions of an electron volt), but it diminishes significantly with increasing neutron energy to be rather smaller (by a factor of 1,000 or more) for neutrons in the MeV range. To take advantage of the higher probability of interaction at low energy some detectors lower the energy of the neutrons by making them pass through a material with low mass nuclei that slow the neutrons down through collisions, a process called moderation. For neutrons in the MeV range, tens of centimeters of a moderator material such as polyethylene surrounding the detector are required. Some other detectors simply rely on the fast neutrons to induce the reactions, and still others rely on the protons (simple hydrogen nuclei) that recoil from collisions between energetic neutrons and hydrogen atoms in the detector. The relatively low efficiency of all of these detectors at high neutron energies make fast neutrons less attractive for signal transmission (see Knoll, 2000).

FINDING

Finding: A variety of accelerator systems is available which can generate electron beams and x-ray radiation at different energies and with different output powers. Accelerators are widely used in industry and are theoretically able to replace radionuclide sources in almost any application, although practical and economic factors limit their value as replacements for some applications.

CHAPTER 5

SELF-CONTAINED IRRADIATORS

SUMMARY

Self-contained irradiators[1] are used mostly for blood irradiation, biomedical and radiation research, and calibration of other devices. Blood banks irradiate selected units of blood to prevent at-risk patients from developing graft-versus-host disease (GVHD), a rare but usually fatal complication of transfusion. Most self-contained irradiators use a radionuclide radiation source, with the vast majority being cesium-137 because of its long half-life, relatively low cost, and relatively modest shielding requirements, which make it possible to place a device in the upper floors of some hospitals and blood banks. Several alternatives to the use of cesium-137 for blood irradiation currently exist, including cobalt-60 sources, which, unlike cesium-137 sources, would be resistant to dispersal, as well as a commercially available x-ray device and existing hospital linacs, which do not contain radionuclide sources. All existing replacements are currently more costly than the current cost of purchasing and operating cesium-137 blood irradiators. In addition, if high-activity glass or pollucite or other ceramic cesium-137 sources were available, then new cesium-137 irradiators could be designed to use them, and some old irradiators could even be retrofitted, depending on the size and shape of the new source pencils. These alternative forms could reduce the devices' hazard with respect to malevolent uses.

Another kind of self-contained irradiator is a calibration chamber for high-dose-rate radiation detectors used by nuclear power plants and some other facilities. The calibration irradiators typically use about 15 TBq (400 Ci) but some hold as much as 82 TBq (2,200 Ci) of cesium-137. Some of these calibrators are located at nuclear power plant facilities, which are under more stringent security requirements than facilities with only materials licenses, but over 100 are located outside of nuclear power plants. The following sections describe blood irradiation, research irradiation, and calibration laboratories, including what they are needed for and whether replacements are technically feasible.

BLOOD IRRADIATION

Hospitals and blood banks irradiate blood products to prevent transfusion-associated graft-versus-host-disease (GVHD). Transfusion-associated GVHD is a deadly transfusion complication resulting when some donor white blood cells (specifically, T lymphocytes, Figure 5-1)

[1] Self-contained irradiators are classified as "Category I" irradiators by the American National Standards Institute (ANSI), ANSI Standard N433.1, Safe Design and Use of Self-Contained, Dry Source Storage Gamma Irradiators (Category I). This is not to be confused with the IAEA Category 1 definition. This ANSI standard defines a Category I irradiator as "[a]n irradiator in which the sealed source(s) is completely contained in a dry container constructed of solid materials, the sealed source(s) is shielded at all times, and human access to the sealed source(s) and the volume(s) undergoing irradiation is not physically possible in its designed configuration." The dry storage is to distinguish the self-contained irradiators from the large panoramic sterilization irradiators, which are stored in a water pool.

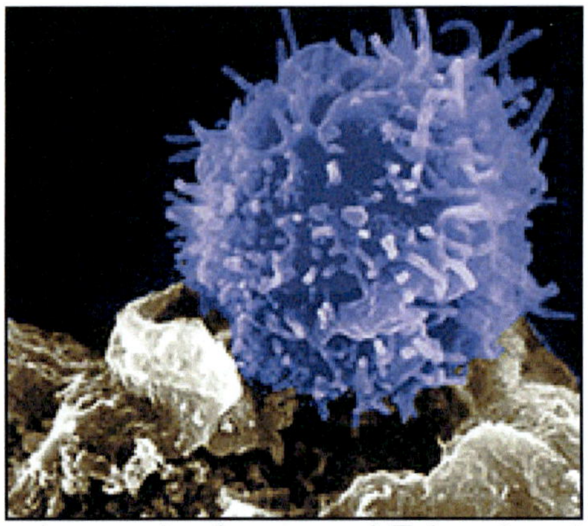

FIGURE 5-1 Scanning electron micrograph of a normal T lymphocyte. SOURCE: Image courtesy of Lawrence Berkeley National Laboratory (Yaris, 2003).

attack the recipient's tissues.[2] GVHD can develop either when the recipient's immune system does not recognize the donor's foreign white blood cells as different or when the recipient's immune system is weakened or defective and is unable to contain and eliminate transfused lymphocytes. Chemotherapy or radiotherapy treatments are prescribed for patients with a number of hematological cancers requiring bone marrow or peripheral blood stem cell transplants. Donor T cells that contaminate red cell and platelet units will recognize the recipients' cells as foreign, proliferate, and mount an immune response against recipient tissues. In approximately 90 percent of the cases, this immune response, or transfusion-associated GVHD, leads to tissue destruction, organ failure, and death (Butch, 1996).

To prevent GVHD in transfusion recipients, blood banks typically irradiate red blood cells and platelet components. Irradiation produces ionizations and free radicals that damage genetic material in the white blood cells and inhibit cell division (replication) (Plappert et al., 1995; van Ankeren et al., 1988). T cells require cell division to mount an effective immune response. Therefore, treatment of blood with sufficient ionizing radiation prevents T-cell proliferation and GVHD. Neither red cells nor platelets contain the genetic material for replication, and damage to genetic material in platelets (mitochondrial DNA) does not compromise the platelet's use. Irradiation of blood with a dose of 25 Gy delivered from a blood irradiator or a radiotherapy linear accelerator (linac) to blood containers kills 99.9995 percent (a reduction factor of 200,000) of white blood cells that contaminate red cell products and more than 99.9988 percent (a reduction factor of 80,000) of white blood cells that contaminate platelet products, respectively, so that less than approximately 1/100,000th of the initial white cells are viable.

Current guidance from the Food and Drug Administration (FDA) recommends a dose of 25 Gy delivered to the midplane of the blood container with no part of the blood container to receive less than 15 Gy. This dose is sufficient to prevent GVHD and has no impact on platelet properties, shelf life, or in vivo platelet recovery and survival, and minor effects on red cell properties, shelf life, and in vivo circulatory recovery and survival (Butch, 1996). Irradiators are dose mapped once per year, and times for exposure are calculated either once (for cesium-137 sources) or four times (for cobalt-60 sources) per year to ensure that doses delivered to blood products meet these quality control standards (Luban et al., 2000; Moroff and Luban, 1997; Pelszynski et al., 1994).

Approximately 10.5 percent of blood components that were produced in the United States (of about 15.5 billion total units prepared) in 2001 were irradiated (Sullivan et al., 2007). It is estimated that only about one-fifth of all irradiated red cell units are produced at centralized

---

[2] T lymphocytes are specialized white blood cells that identify and destroy invading organisms such as bacteria and viruses.

blood centers; the remaining units are irradiated at hospitals. The higher production of irradiated red cell units by hospitals may be due to the shortened shelf-life of gamma-irradiated red cell products (28 days rather than 42 days). In addition, some physicians are concerned about infusing irradiated units that have been stored for several days or weeks to infants and some adults. This is because irradiated red cells leak potassium at twice the rate of normal red cells and some patients are susceptible to adverse cardiac events from potassium following transfusion. Therefore, in many hospitals, red cell units are irradiated just prior to transfusion. An alternative explanation for high hospital production of irradiated products is that most hospitals already have the equipment for irradiation and thus do not wish to pay fees to blood centers for this service. Much greater than 10 percent of the 10.3 million platelet units produced in 2001 are estimated to have been gamma irradiated (Sullivan et al., 2007).

From 2002 to 2004, the use of bone marrow and peripheral blood stem cell transplants increased approximately 8.8 percent annually. In recent years, experimental transplant activity has expanded to include not only patients with hematological cancers, but also those with genetic abnormalities in their hemoglobin (hemaglobinopathies), such as sickle-cell anemia, thalassemia, and autoimmune diseases such as scleroderma, multiple sclerosis, rheumatoid arthritis, juvenile idiopathic arthritis, and systemic lupus erythematosus. Nontransplant patients requiring transfusion, including patients with some congenital immunodeficiencies, immunodeficiencies from viral infection, and low-birthweight or in-utero neonates with immature immune systems, may also require gamma-irradiated blood products (McCullough et al., 2006; van Laar and Tyndal, 2006).

Aside from irradiation, there are other ways to inactivate white cells or one can remove them from red cell and platelet components. Investigating these alternatives is, however, difficult to do in the United States because there is an ethical dilemma in conducting a trial to investigate use of an experimental prophylactic treatment to prevent a deadly disease when a well-known, efficacious alternative (irradiation) is readily available.[3] The existing alternatives are described below.

Cesium Blood Irradiators

A simplified design of a typical cesium-137 blood irradiator is shown in the schematic in Figure 5-2, a photograph of a typical blood irradiator is shown in Figure 5-3, and the technical specifications for four irradiators are given in Table 5-1.

One to six blood containers (usually bags) are labeled for quality control and placed in a metal canister by an operator who then loads the canister into the irradiator. After automatically shielding the operator from source exposure and simultaneously positioning the canister adjacent to one or more shielded pencil-shaped cesium-137 sources, the irradiator rotates the canister to ensure more uniform gamma-ray exposure of the entire volume of each blood unit.

---

[3] The potential for GVHD depends on the population (it is higher in transplant recipients and in countries where genetic variation is small) and so, selection of the study population would determine the study size. In a more susceptible population, the size of the trial would need to be smaller than in a less susceptible population.

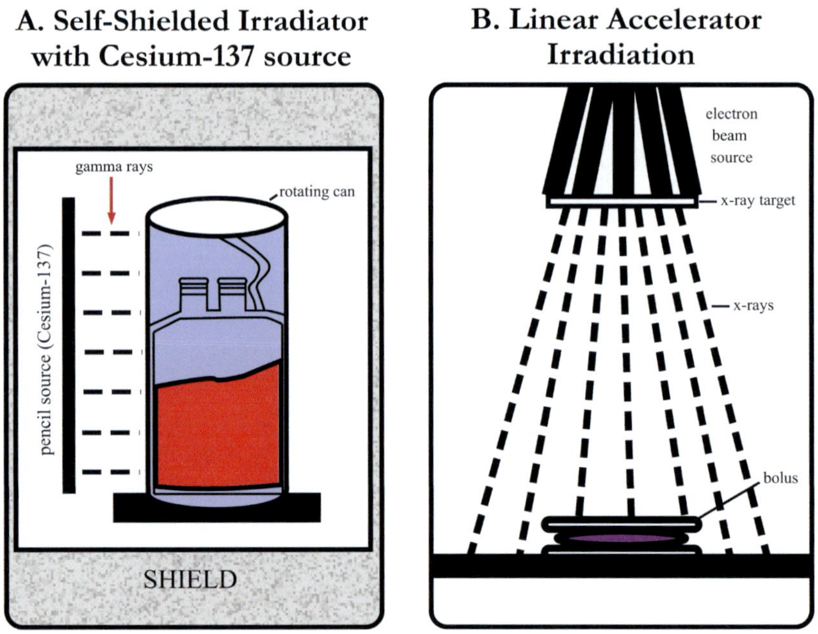

FIGURE 5-2 Diagrams of (A) a configuration of a gamma irradiator using cesium-137 and (B) a configuration of a linac irradiator. The plastic bolus is a container that enhances the dose uniformity in the irradiation configuration shown. SOURCE: Image provided by the committee.

FIGURE 5-3 Typical self-shielded blood irradiator and typical 500-ml blood bag and small irradiation canister. SOURCE: Images provided by the committee.

TABLE 5-1 Comparison of Cesium-137 Irradiators

	Model			
Specification	**Nordion** GC 1000 Elite	**Nordion** GC 3000	**CIS** IBL 437C	**Shepherd** 143
Load (kg/sq m)	1,467	1,886	5,200	Not available
Weight (kg)	1,150	1,479	2,150	907 or 1,814
Height (m)	1.55	1.55	1.50	2
Width (m)	0.8	0.8	0.67	0.6
Depth (m)	0.98	0.98	0.650	0.6
Outlet (V)	110	110	100–240	110
Activity (TBq) [Ci]	24.1–107.4 [575–2,900]	53.7 or 107.4 [1,450 or 2,900]	63–189 [1,700-5,100]	42.8–259 [1,155–7,000]
Canister (liters)	0.824	2.34	3.8	0.6–3.9
Time to deliver 25 Gy (min)	1.6–7.14	2.56–5	2.8 or more	Not available

SOURCE: Courtesy of Gammacell® 1000 Elite/3000 Elan (MDS Nordion, 2006); CIS IBL 437C (CIS-US, Inc., 2007); Shepherd 143 adapted from Cook (1996), with additional information from the Sealed Source and Device Registry.

An example of a dose map of one canister is given in Figure 5-4. If the dose at the midplane is defined as 100 percent, the dose in any portion of the canister other than the very top and bottom center portion does not typically vary by more than ±20 percent. Exposure times are typically several minutes; irradiators containing higher activity sources have shorter exposure times due to their increased dose rate. An irradiator takes twice as long to deliver a given exposure after the source has decayed through one half-life. Given the 30.2-year half-life of cesium-137, a typical blood irradiator is in service for approximately 30 years before needing to reload the source, or more commonly replace the entire irradiator to maintain a practical exposure time. Because blood irradiators are self-contained (they have built-in shielding), they need not be located in a bunker in the basement of a hospital or blood center. The weights of cesium blood irradiators do, however, require more support than is found in the upper floors of many buildings, in which case the irradiators may need to be located next to a structural support for the building or have additional support installed to spread out the weight.

REPLACEMENTS FOR BLOOD IRRADIATORS

There are several alternatives to radioactive cesium chloride self-contained irradiators. These include different material forms for cesium-137 sources, cobalt-60 sources, x-ray sources, chemical inactivation, and filtration techniques. Each of these is discussed below. The committee spoke with the major self-contained-irradiator manufacturers and found them reluctant to shift toward replacements for radionuclide radiation sources without some incentives.

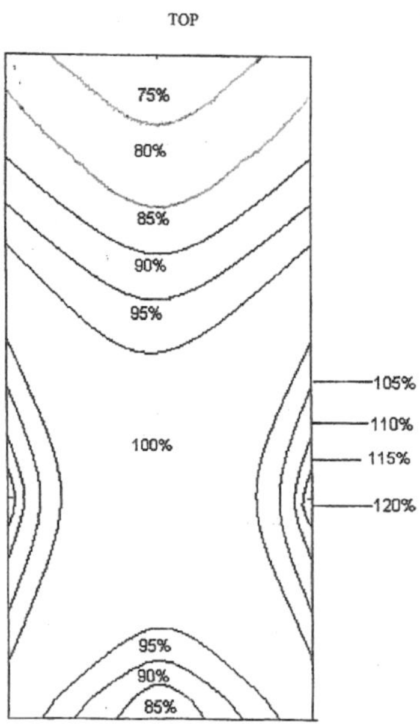

FIGURE 5-4 Typical dose map in a blood irradiation canister. SOURCE: Unpublished data on isodose curve in percentages for Gammacell 1000, provided by the American Red Cross, 2007.

Different Material Forms for Cesium-137 Sources

Manufacturers of self-contained radionuclide source devices have commented in the committee's public meetings that the decrease in specific activity from preparing a cesium-containing ceramic would make it impossible to produce current dose rates in existing devices. They have also said that it is likely that new designs of blood irradiators would have to be fashioned to use ceramic or other complexed forms of cesium in a 24- to 100-TBq device.

Increasing the number of pencil sources by a factor of two to five would be expected to produce dose rates from cesium pollucite irradiators that are equivalent to existing cesium chloride irradiators. The committee examined several designs of self-contained irradiators, and some of them can accommodate several different source loadings with little change in the rest of the design. This is possible because the shielding is designed for the maximum activity loading and the source holder can accommodate more than one source pencil. If alternative forms of high-activity cesium-137 source can be produced with cesium densities within a factor of two of the current radioactive cesium chloride sources, then at least some of the self-contained irradiator designs could simply switch to the alternative forms by loading two pencils instead of one, with no retrofits required and no degradation in performance. (Regulatory reviews are needed for alternative forms of radioactive material used in current radiation source devices, as noted in Chapter 10.)

Cobalt-60 Blood Irradiators

Use of blood irradiators containing a radionuclide metal, with limited solubility and considerable resistance to explosive dispersion, would reduce the potential for widespread contamination of large areas if it were used in a radiological dispersal device. There are blood irradiators with the radionuclide metal cobalt-60 instead of the radioactive cesium chloride source. In the late 1980s there was an interruption of fabrication of cesium-137 irradiators because of the closure of a major cesium-137 production facility. During that period, at least one supplier of blood irradiators developed cobalt-60 blood product irradiators. Unlike the cesium-137 blood irradiators in which radioactive pencil source(s) are positioned on one side or in a U shape around a central, rotating canister on a turntable, the cobalt-60 irradiator had multiple fixed sources arranged around a central cavity into which the canister was lowered. These units were stocked with sufficient cobalt-60 that reloading of the radionuclide was not required for 10 to 15 years. Cobalt-60, with its more energetic gamma emission, requires an approximately fourfold increase in the mass of lead shielding compared to cesium-137 for the same activity source. A cobalt-60 source gives a much higher dose rate to the object being irradiated than does a cesium-137 source of the same activity, so a lower activity cobalt source can be used, even if the irradiator is loaded with extra activity to compensate somewhat for cobalt-60's faster decay rate. As a result, the cobalt-60 irradiator units (mainly their shields) weighed only about twice as much as most cesium-137 blood irradiators; a device weighed approximately 2,700 kg. Because the weight required bottom-floor installation in most buildings and because its shortened useful life (15 instead of 30 years) made operation more costly, very few of these irradiators were sold to blood banks and hospitals. However, the use of cobalt may still be viable in many cases. Currently, 31 cobalt-60 blood irradiators are reported to be located in the United States.

X-ray Irradiators

An x-ray blood irradiator originally developed and distributed by Rad Source Technologies, Inc. (Alpharetta, GA), the RS 3000, received Food and Drug Administration approval for use and has been available as an alternative to cesium-137 and cobalt-60 irradiators since August 1999. The device, which is similar in size and lower in weight than cesium-137 blood irradiators, utilizes 160-kVp x rays and can irradiate up to 2–3 units of red cells, depending on the bag manufacturer, with 25–37.6 Gy in approximately 5 minutes (Table 5-2). In one side-by-side study, similar lymphocyte inactivation and red cell potassium release were observed in red cell units treated with this x-ray device compared to units treated with the same dose of cesium-137-generated gamma rays (Janatpour et al., 2005). X-ray–treated red cells exhibited an enhanced degree of hemolysis (breakdown of red blood cells) during storage compared to cesium-137 gamma-ray–treated units; however, these differences were small and not clinically significant.

There are reports that the early RS 3000 irradiators suffered from reliability and service-related problems. In May 2002, the company issued a voluntary nationwide recall of all 20 of its installed RS 3000 blood irradiators to complete a cooling system retrofit to prevent overheating and failure. During the retrofit or failure of the irradiator, some users were forced to send blood units to other institutions for irradiation with radionuclide sources. In 2003, the company licensed its x-ray blood irradiator to Nordion, who markets the device as the Raycell®. Because of the history of x-ray device breakdown and the critical nature of providing irradiated blood to patients, there are concerns among some in the blood bank community that radionuclide “backups” of x-ray devices are needed at institutions to be available during times of x-ray device failures.

The Raycell® has achieved a small but growing market penetration. By October 2006, MDS Nordion estimated that there were approximately 100 Raycell® units in operation (United States, Sweden, Germany, France, Italy). Approximately 80 additional x-ray blood irradiators developed primarily by two other foreign manufacturers are in use outside the United States.

Costs of purchasing a Raycell® are roughly comparable to purchasing cesium-137 irradiators, based on user information provided to the committee. Direct operating costs would be expected to be somewhat larger than those of a radionuclide blood irradiator because of the increased costs of electricity in the x-ray device and increased need for service and part replacement. A maximum electricity cost of less than $0.18 for irradiating three units of blood can be estimated by utilizing the stated voltage specification, maximum amperage, 5–minute irradiation time (Table 5-2) and assuming $0.12/kWhr for electricity costs. A course for training service personnel for installation and maintenance of the Raycell®, including x-ray tube replacement and alignment, high-voltage generator replacement, and dosimetry is offered by MDS Nordion and listed as Can$1,500. However, service needs for a failed instrument are urgent, the complexity of required preventive maintenance is much greater than for radionuclide sources and beyond the abilities of many blood bank staff, and the need to maintain good manufacturing practice conditions lead users to purchase an annual maintenance and service agreement for the device. The current cost of an agreement is approximately $10,600 per device per annum, and costs are expected to increase as instruments age. Thus, over a 30-year period, or the expected lifetime of a cesium-137 blood irradiator, use of the Raycell® x-ray device instead of the radionuclide irradiator is expected to incur at least $318,000 in additional service and maintenance costs. The cost of these services therefore increases the cost of irradiating blood by 177% over the purchase cost ($180,000, as reported by MDS Nordion) of a new cesium-137 irradiator, which are reliable in the field and require less complex maintenance. The committee contacted some organizations that currently own and operate Raycell® devices, and despite this additional cost, they were actually considering purchasing more of the x-ray devices rather than the gamma irradiators. One factor may be that a blood irradiator processes many blood units over its lifetime, so the cost difference on a per-unit basis could be relatively small.

In addition, the costs of disposal or retirement of an existing cesium-137 blood irradiator that an owner would incur with a transition to x-ray irradiators is estimated by manufacturers (MDS Nordion, JL Shepherd & Associates) to be $35,000 to $40,000. These costs are primarily for transportation of sources. There is currently no disposal facility available for civilian high-activity cesium sources in the United States. The only options for disposition of an unwanted source are to return the source to the manufacturer for recycling or to request that the source be taken by the National Nuclear Security Administration's Offsite Source Recovery Project (OSRP), which stores the sources free of charge to the companies. Therefore, the actual cost to the country of disposal of high-activity cesium-137 devices is currently unknown. Further, companies and users that rely on continued take-back practices of other companies, and even from the OSRP, are taking risks in light of the potential for shifts in business practices and the sometimes shaky federal funding for the OSRP.

TABLE 5-2 Features of the Raycell® Irradiator

X-ray potential	160 kV
Weight	710 kg
Floor loading	1,115 kg/m^2
Canister dimensions	6 in. × 4 in. (diameter × depth) (15 cm × 9.5 cm), 1.68 liters
Unit dimensions	59.45 in. (151 cm) high, 44.63 in. (114 cm) wide, 21.75 in. (56 cm) deep
Utility requirements	200–240 V ac, 50/60 Hz single phase, 60 A maximum, 380–440 V ac, 50/60 Hz three phase, 40 A maximum
Water flow rate and pressure	2.6 gpm (10 L/min) 50–70 psi (345–483 kPa)
Time to deliver 25 Gy (min)	~5

SOURCE: Courtesy of MDS Nordion (2006).

Use of the Raycell® instead of a radionuclide blood irradiator would be expected to have reduced indirect operating costs, because x-ray devices are not subject to the increased security controls required by the U.S. Nuclear Regulatory Commission (U.S. NRC) for gamma irradiators and other Category 1 and 2 devices (see Chapter 3), and also may have fewer other regulatory burdens because the device contains no radionuclides. However, not all savings in security costs would necessarily be realized if for example, the facility has other reasons to employ security personnel.

Other kilovoltage sources will soon enter the market, produced by manufacturers of research irradiators. See the section below on Research Irradiators.

Electron-Beam Irradiators

Electron beams have been used to sterilize food and low-density medical products (Kunstad, 2001) in devices similar to large-scale x-ray irradiators but lacking high-density targets used to generate x-rays. Because electron-beam irradiators do not need to perform the inefficient conversion of electron to x-ray energy, they do not waste 90 to 95 percent of the electrical power needed to supply these devices. In beam generators, 5 or 10 MeV electrons directly bombard the product to be sterilized as it is passed through the beam. For food products with a density of approximately 1 g/cm^3, electron penetration is limited to approximately 5 cm. Based on the fact that blood products have densities similar to those of food products, it is expected that only one blood product thickness (about 2 to 4 cm) could be irradiated at a time.

No electron-beam irradiators have been used to routinely treat blood products to prevent GVHD, although the doses necessary for prevention of GVHD are roughly 1,000-fold less than those needed to sterilize other types of products. An electron-beam food irradiator has capital equipment costs up to $3.5 million, with a large cost associated with the waveguide of the instrument. With high capital costs, one could consider shipping blood products to a centralized electron-beam irradiator. However, it is not feasible to have centralized electron-beam irradiation of blood. Blood products typically are irradiated "on demand." The short 5-day shelf life of platelet products, in particular, would not permit adequate shipping time, "in queue" time as other materials are irradiated, and processing time necessary to send units to centralized irradiation facility. The development of small on-demand electron-beam devices specifically designed for blood irradiation has not been explored, and its development, as the state of the art exists today, would be hampered by the high cost of the instrument and in particular the cost of the waveguide. To be competitive with radionuclide blood irradiators, an e-beam instrument would have to be developed that could be marketed for approximately $200,000 per device. Very low energy electron sources up to 150 kVp have been developed in this price range, but the penetration of such low energy electrons is insufficient to be of use in this application.

Linacs

Linacs that are routinely used for radiotherapy of cancer patients have also been successfully used for blood irradiation (Moroff and Luban, 1997). With their high dose rates, large radiation field (up 40 cm × 40 cm) and x-ray energies capable of delivering a uniform dose in a 25-cm-thick volume when opposed beams are used, many units of blood can be simultaneously irradiated within a 5- to 10-minute period. Because their capital costs are already supported by radiotherapy departments, and patient use is usually limited to the daytime hours, evening use of linacs for blood irradiation is feasible. Irradiation would need to be carried out by a trained radiation therapist or physicist, rather than blood bank staff, so operating costs may be greater than those associated with gamma irradiation of blood using a radionuclide source. The

Indiana Blood Center, which uses an x-ray irradiator to irradiate blood for several hospitals, told the committee that it charges $55 per unit for irradiation services that otherwise would be done using the hospital's in-house linacs for irradiation. Further, blood irradiation by a hospital department other than the blood bank may be inconvenient because of the need to accommodate the radiation oncology schedule. In addition, the high capital costs of acquiring linacs would be prohibitive for use at nonhospital blood centers. Hospital-based linacs could serve as a viable backup if a blood irradiator became dysfunctional, and in some circumstances, could replace cesium-137 based hospital blood irradiators.

Nonirradiation Approaches

Leukoreduction

Filters to diminish white cell levels in red cell and platelet units by a factor of 1,000 to 100,000 (3 $\log_{10}$ to 5 $\log_{10}$) are currently available and are routinely used in a majority of cellular blood products. In theory, a reduction in donor leukocyte counts would be expected to protect against GVHD. However, several cases have been reported documenting the occurrence of GVHD despite leukoreduction with previous generation filters that were not as efficient at removing lymphocytes as are current filters (Garcia Gala et al., 1993; Hayashi et al., 1993; Heim et al., 1992). Because of these reports, filtration of blood with the current generation of filters generally is not considered by physicians to be effective prophylaxis against GVHD, although it is actually unknown whether current high-efficiency (5 $\log_{10}$) filters might be an effective substitute for gamma irradiation. Even if 5 $\log_{10}$ leukoreduction could be demonstrated to prevent GVHD, filter breakthough, or filtrations in which the substrates fail to retain the necessary amount of leukocytes to diminish white cell levels by a factor of 1,000 to 1000,000, of leukocytes is estimated to occur at a frequency of 1 in 500. Whether the extent of breakthrough would be sufficient to promote a GVHD reaction is also unknown.

Pathogen Reduction Techniques

Several methods to inactivate viruses and bacteria in blood components have also shown promise in inactivating white cells at levels that may be useful in prophylaxis against GVHD. All the techniques involve the use of nucleic acid damaging agents. Two existing techniques utilize ultraviolet A (UVA) light to induce photochemical reactions in the agents that attach themselves to nucleic acids critical to reproduction of viruses, bacteria, parasites, and white cells. Amotosalen (S-59), developed by Cerus Corporation, is a synthetic psoralen that inactivates pathogens through such phototreatment. Following phototreatment, platelets containing amotosalen and photoproducts are transferred to a container with a resin designed to reduce the concentration of drug and photoproducts.

Studies with amotosalen and UVA light have demonstrated pathogen reduction and inactivation of human white cells comparable to and greater than that achieved in irradiation (a factor of 250,000 reduction in viable white blood cells) (Grass et al., 1998). This technique has been effective in an animal model and three small clinical trials in Europe (Corash and Lin, 2004; Grass et al., 1998). Cerus is licensed in Europe for sale of their amotosalen and UVA light irradiation system for treatment of apheresis platelet units and buffy-coat platelet pools. Costs of the system are estimated to be $80 to $100 per unit.

In a system under development by Gambro BCT Corporation, riboflavin (vitamin B_2) is used with UVA light treatment to inactivate viruses, bacteria, parasites, and white cells in apheresis and buffy-coat-derived platelet components (Hardwick et al., 2004; Ennever and

Speck, 1981). Because riboflavin is a substance that is generally recognized as safe[4] and the phototreatment product, lumichrome, is not considered to be toxic, riboflavin and photoproducts are not removed in any post-phototreatment step (Ennever and Speck, 1981). A study suggests that this technique may be effective for inhibition of lymphocyte proliferation (Fast et al., 2006). The study demonstrated inactivation of white blood cells up to the limit of detection of the assays used, which is a factor of about 100. A more sensitive assay would be needed to demonstrate whether this method is effective in preventing GVHD, which requires orders of magnitude greater reduction.

Phototreatment with both amotosalen and riboflavin result in a smaller fraction of treated platelet products being retained and recirculated in the body compared with using gamma irradiation (AuBuchon et al., 2005; McCullough et al., 2004; Snyder et al., 2004). Thus, both phototreatment alternatives would require the use of more treated platelets than does gamma irradiation. Because of high UV absorbtion in red blood cells, these approaches have not been used for whole blood.

Cerus Corporation has also developed a DNA-specific alkylating agent, S-303, to inactivate viruses, bacteria, and parasites in red cell components. Although there are no published data reporting inactivation of white cells, it is anticipated that S-303 will be effective on white cells based on robust inactivation of bacteria and viruses, which have nucleic acid target cross sections that are 10–10,000 times smaller than those of mammalian cells.

S-303 also reacts with nucleic acids, but without the UVA light trigger. S-303 is designed to break down in blood to reduce the genotoxicity (mutagenicity and potential carcinogenicity) and general toxicity of the compound, and is used with reduced glutathione, a so-called quencher, to reduce reactivity of S-303 with the red cell membrane. Following incubation of red cells with S-303 for 24 hours at room temperature, red cells containing the S-303 reaction product are transferred to a container with a resin designed to reduce the concentration of either S-303 or its reaction product, although based on its lifetime of approximately 20 minutes none is expected to remain.

In Phase I and II clinical trials, the 24-hour recovery and survival of red cells that have been treated with S-303, stored for 35 days, and infused into normal autologous donors is comparable with that of normal donors receiving autologous untreated and similarly stored red cells (Hambleton et al., 1999; Cook et al., 1998). However, in a Phase III chronic transfusion study, three patients developed antibodies to S-303–treated red cells (Benjamin et al., 2005). In addition, in vitro studies on compounds with structures similar to S-303 demonstrated alkylations to proteins, including residual alkylations to the red cell surface despite the addition of reduced glutathione to prevent red cell modification (Cook and Stassinopoulos, 1998; Creech and O'Connell, 1981; Lauffer et al., 1979). A new methodology has been developed to reduce the amount of S-303 reacting with the red cell surface. Further development of S-303 will require repeating clinical studies with revised protocols to prevent red cell surface modifications.

RESEARCH IRRADIATORS

Research irradiators are used to expose biologic and nonbiologic materials to radiation of various types in order to evaluate the response of target materials to various doses, dose rates, and energies of the applied radiation source. Such units are used in a limited way in

---

[4] Note that "generally recognized as safe" (GRAS) is a technical term: "under sections 201(s) and 409 of the Federal Food, Drug, and Cosmetic Act (the Act), any substance that is intentionally added to food is a food additive, that is subject to premarket review and approval by FDA, unless the substance is generally recognized, among qualified experts, as having been adequately shown to be safe under the conditions of its intended use, or unless the use of the substance is otherwise excluded from the definition of a food additive" (FDA, 2004).

materials research and extensively in radiobiologic research. They are used to evaluate electronics components and satellite components as well.

Radiobiologic research involves either the exposure of bacterial, yeast, or mammalian cells to graded doses of radiation in order to evaluate response or the exposure of whole animals or portions of live animals in order to evaluate the response versus dose. Biologic exposure may also be a tool to enable other studies to be done, such as causing immunosupression so that transplantation may be evaluated. Research irradiation has been done with two primary types of irradiators: beam units located in a shielded room and self-contained irradiators with built-in shielding. The beam units are similar to radiotherapy cesium or cobalt units but are located in a shielded room in a research laboratory. They will deliver dose rates of 1 to 3 Gy/min at 50- to 80-cm distance.

Self-contained units are housed in a dedicated room in a laboratory (see Figure 5-5). Often some of the units will be located inside an animal facility to allow irradiation of pathogen-free mice without removing them from the protected clean facility. The self-contained units have a cavity large enough to allow placement of partial body shields for small animals and dose-rate modifying shields. Units must deliver dose rates of 1 to 10 Gy/min to a cavity of 4 to 10 liters in size. Ideally the energy is high enough to make the effect similar in relative biologic effectiveness (RBE) to x-rays of 1 MeV or higher in energy, so that RBE corrections do not need to be made. These units can be loaded with cesium-137 or cobalt-60. Required shielding is two times thicker for cobalt units.

Research irradiation facilities designed to study biological effects from continuous low-dose-rate exposure over periods of days, weeks, or more are not feasible using conventional x-ray machines or accelerators. No current compact x-ray sources can operate continuously and steadily for such long time periods.

Worker hazards are minimal with these units as long as adequate door and source location interlocks are functional and shielding is of adequate thickness. Cost of the units depends on the cavity size, the radionuclide used, and the dose rate desired. Prices range from $150,000 to $500,000. In addition, one must consider the cost of security for the sources and cost of disposal of the sources when decayed or no longer used.

Security must prevent access to the unit and removal of the sources. In June 2006, the U.S. NRC and the Agreement States imposed increased controls on irradiators that contain more than threshold quantities of radioactive material (see Chapter 3). Typically, research irradiators are located in facilities that have additional security, particularly if located in animal care units where security is in place for reasons unrelated to the irradiator.

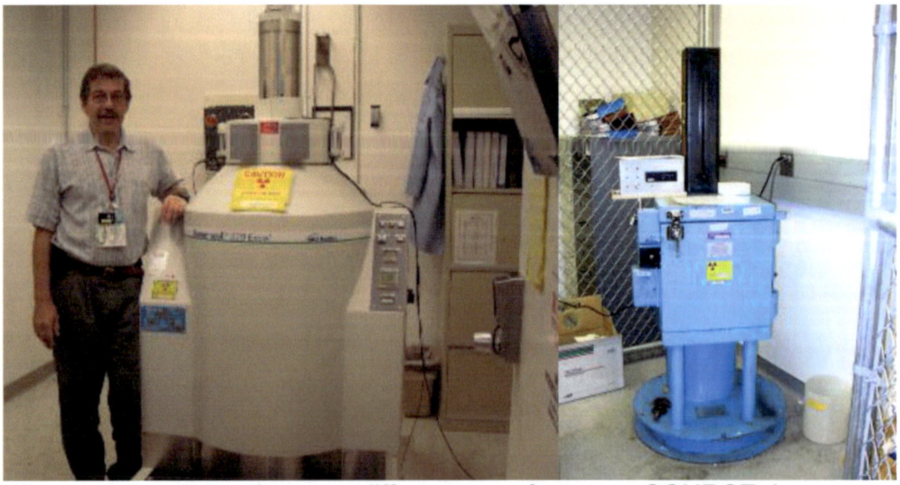

FIGURE 5-5 Research irradiators from two different manufacturers. SOURCE: Images provided by the committee.

Replacement Technologies

Prior to the wide availability of radionuclde irradiators, x-ray sources were widely utilized in research applications. Usually a kilovoltage (200–300 kVp) radiotherapy unit is used in a shielded room or located in a shielding box. In some cases for cell irradiation, units with energy as low as 50 kVp are employed. These units provide dose rates of 0.5 to 1.5 Gy/min depending on the target-to-surface distance. Some of these units are still in use.

Currently at least two companies market modern kilovoltage systems for specimen irradiation. Precision X-ray, Inc., sells a series of x-ray irradiators ranging in energy from 160 kVp to 320 kVp with an exposure chamber large enough for some animals. Units have been sold to over 30 institutions, some having up to 5 units each. Output ranges from 100 to 3 Gy/min depending on filtration and source-to-surface distance (see Figure 5-6a). Another company, Rad Source Technologies, Inc., has developed a center-filament x-ray tube that irradiates 360 degrees around the tube. This new tube has a cylindrical gold target that will be used in a new type of specimen and blood irradiator that Rad Source reports will be capable of up to 450 Gy/min with multiple tubes and rotating specimen chambers (see Figure 5-6b). As discussed previously, Rad Source Technologies developed the Raycell® now produced by MDS Nordion.

These types of devices have the advantage of large fields of irradiation and the freedom from radionuclides and their security and disposal costs. These standard kilovoltage units are reliable, often lasting 30 years. Tube life is estimated at 10 years, and replacement cost is about $18,000. The Rad source unit tube can be returned to the factory and a new filament installed in the tube. Costs of these units range from $120,000 to $150,000, higher for the very high output devices. Both of these companies plan dedicated blood irradiation systems to be marketed in the near future. The disadvantage of such units in radiobiologic research is the kilovoltage x-ray energy which exhibits an increased RBE, although one can correct for this. Megavoltage linacs can also be used, usually nights and weekends in the radiotherapy department.

(a) (b)

FIGURE 5-6 X-ray research irradiators from (a) Precision X-ray (X-RAD 320) and (b) Rad Source Technologies (RS 2500). SOURCE: Images courtesy of (a) Precision X-ray, Inc., and (b) Rad Source Technologies, Inc.

It is possible to design higher energy x-ray units with high output that could be employed for this purpose (see Chapter 4). Such units could replace radionuclide units and would require more shielding and be more expensive than the kilovoltage units discussed above.

All of the cesium research irradiators could be replaced by cobalt-60 units, the sources for which are readily available and have potentially higher output and larger fields of irradiation. Because cobalt sources need more frequent source replacement, the acquisition, transportation, and source replacement costs would be higher for operation, but disposal of spent sources is available. They could also be replaced with kilovoltage x-ray units similar to those described above or with small linacs in shielded rooms.

CALIBRATION SYSTEMS

Calibration systems use high-activity radiation sources (approximately 15 to 82 TBq [400 to 2,200 Ci]) to produce radiation fields of known intensity for calibration of radiation monitoring equipment and dosimeters, whereby the equipment and dosimeters can be evaluated for accurate operation. A source of measured activity is required to calibrate instruments and dosimeters to accepted standards. Figure 5-7 shows a diagram and a photograph of a typical gamma-beam calibration source.

The system usually consists of radioactive sources, radiation shielding, a mechanism for positioning the source, and a track or internal chamber for positioning the items to be calibrated. Modern calibration systems may contain a computer controller and safety systems, such as video monitoring, radiation monitors, warning lights and indicators, and a safety interlock system. Although calibration systems may contain different sources for the calibration of gamma, neutron, and beta monitoring equipment and dosimeters, the typical Category 2 sources used for calibration of beta/gamma survey instruments and dosimeters are strontium-90, cesium-137, and cobalt-60. The U.S. NRC Interim Inventory (2007a) reports 104 calibration irradiators using Category 2 sources in the United States, in addition to calibration irradiators at nuclear power plants. These are primarily located in commercial and government calibration facilities and state regulatory agencies. Additional security is required at most of these facilities because of other nuclear material or radioactive sources that are used at the facilities or for other reasons. Replacement of the cesium chloride sources could be made with glass or pollucite forms of cesium since very high specific activity is not required.

According to contemporary national and international radiation dosimetry protocols, Primary Standards Dosimetry Laboratories (PSDLs) and Accredited Dosimetry Calibration Laboratories (ADCLs) are required to provide users' ionization chambers with calibration coefficients obtained in cobalt-60 gamma-ray beams. Therefore, PSDLs and ADCLs incorporate cobalt-60 irradiators, usually decommissioned clinical teletherapy machines, with cobalt-60 teletherapy sources with an activity of 50 to 370 TBq (1,500 Ci to 10,000 Ci).

In the United States, the National Institute of Standards and Technology (NIST) in Washington, D.C., serves as the primary radiation dosimetry laboratory, and there are three accredited dosimetry calibration laboratories.

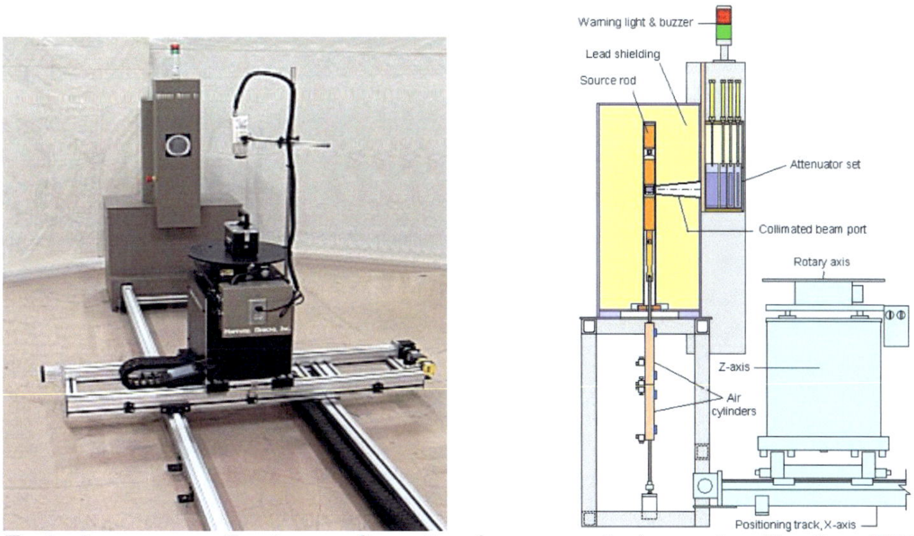

FIGURE 5-7 Typical gamma calibrator configuration for survey instrument calibration. SOURCE: Image provided by Hopewell Designs, Inc. (2007).

SUMMARY AND FINDINGS

In most (and perhaps all) applications discussed in this chapter, radioactive cesium chloride can be replaced by (1) less hazardous forms of radioactive cesium, (2) radioactive cobalt, or (3) nonradionuclide alternatives. However, not all of these alternatives are available now, and all are currently more expensive than radioactive cesium chloride for the users. Use of the more robust but lower specific activity cesium-137 source matrixes may require redesign of some self-contained irradiators, although others might be able to use the new sources without retrofit or any significant change in performance.

Finding: In most (and perhaps all) applications, radioactive cesium chloride can be replaced by (1) less hazardous forms of radioactive cesium, (2) radioactive cobalt, or (3) nonradionuclide alternatives. However, not all of these alternatives are available now, and all are currently more expensive than radioactive cesium chloride for the users.

Some alternatives to radioactive cesium chloride include radioactive cesium glass and a mineral form (pollucite) loaded with radioactive cesium (described in Chapter 2). These alternative material forms use the same cesium-137 as radioactive cesium chloride; thus, the gamma rays and the half-life are identical, but the specific activity of these sources is smaller and the pollucite is more difficult to fabricate, especially for high-activity sources. The committee judges that none of the current applications of high-activity cesium sources about which it was informed requires the higher specific activity afforded by cesium chloride. Accommodating the larger volume needed to achieve the same source activity would require redesign of some (not all) devices. High-activity cesium sources are not, however, available in these alternative material forms today, and making them available may require the cesium source producer (the Production Association Mayak in Russia) to modify its production process.

Cobalt-60 may be substituted for radioactive cesium chloride for many applications (see the discussion in this chapter, on Cobalt-60 Blood Irradiators), although as much as twice the shielding thickness may be required for a source that achieves the same dose rate, and the half-life of cobalt-60 is shorter (5.3 years for cobalt-60 versus 30 years for cesium-137) thus lowering significantly the useful lifetime of the source. Shielding challenges can be addressed in part by switching from lead shields to more effective tungsten or depleted uranium shielding, but tungsten shielding is more expensive than lead and manufacturing depleted uranium shielding is a very specialized, expensive operation. The shorter useful lifetime of radiation sources requires that they be replaced periodically, which entails transportation of a fresh source and, in some cases, the used source, with the attendant risks associated with source transportation.

X-ray generators are already commercially available as substitutes for applications that do not require gamma rays with the definite energies emitted by cesium-137, and cobalt-60. x-ray tubes can be expensive and require more maintenance than radioactive sources for periodic calibration and replacement. There is new innovation in x-ray irradiators by at least two companies and more replacements for radionuclide radiation sources could come with some incentives.

CHAPTER 6

PANORAMIC IRRADIATORS

SUMMARY

Panoramic irradiators are operated on a contract basis to irradiate single-use medical devices and products, cosmetics, food, and plastics. Their largest business comes from sterilization of medical devices and products. To date, only gamma and electron-beam (e-beam) irradiators have operated on a large-scale commercial basis. The first large x-ray facility is expected to come into operation in the next few years, in Europe. Gamma, x-ray, and e-beam irradiation can all be effective for the different products, although there are some advantages to gamma and x-ray irradiation for thick or dense packages and advantages to e-beam for products that demand high doses.

Gamma and x-ray irradiation are nearly interchangeable from a physics perspective (x-ray irradiation can have higher energy and therefore have slightly better penetration), and so, x-ray irradiators could be a direct replacement for gamma irradiators. There are practical differences between the gamma and x-ray facility designs and operations that could result in differences in costs. Whether x-ray irradiators are economically competitive with gamma irradiators is not clear. A crude and somewhat incomplete cost analysis suggests that the costs could be comparable for a high-throughput facility, but the actual cost differences depend on variable factors such as the cost of electricity, the reliability of the equipment in the x-ray facility, the facility configuration, and the products to be irradiated.

Ethylene oxide (EO) also is used for chemical sterilization of some products. Given the accidents and potential security risks, health risks associated with exposure, and pressures to encourage EO users to switch *away* from EO because of its toxicity, it is not clear that a shift from irradiation to EO sterilization would be desirable.

USES OF PANORAMIC IRRADIATORS

Panoramic irradiators[1] or gamma irradiation facilities are used to sterilize medical devices and products. They are also used to sterilize pharmaceuticals and consumer products (e.g., cosmetics); sterilize male insects to inhibit infestations; kill bacteria and fungi and preserve color in foods; and process polymers to achieve specific characteristics, such as increased hardness or durability. However, in the United States the sterilization of medical supplies and

---

[1] Panoramic irradiators are sometimes described by their American National Standard Institute category, defined below. These should not be confused with the IAEA categories of sources.

Category II — Panoramic, dry source storage irradiator. American National Standard N43.10. A controlled human access irradiator in which the sealed source is contained in a dry container constructed of solid materials, and the sealed source is fully shielded when not in use; the sealed source is exposed within a radiation volume that is maintained inaccessible during use by an entry control system.

Category IV — Panoramic, wet source storage irradiator. American National Standard N43.10. A controlled human access irradiator in which the sealed source is contained in a storage pool (usually containing water), and the sealed source is fully shielded when not in use; the sealed source is exposed within a radiation volume that is maintained inaccessible during use by an entry control system.

devices constitutes by far the largest enterprise among these activities. Sterigenics, Inc., estimates that the current medical device radiation sterilization market is approximately 5.7 million cubic meters per year (m^3/yr, or 200 million cubic feet per year, ft^3/yr), with about 80 percent using gamma irradiation and about 20 percent using e-beam irradiation (Smith, 2006). This is probably around half of the entire sterilization market, the rest being carried out using other methods. Whether irradiating flies, food, or syringes, these applications generally require high-throughput irradiation to be economically and/or logistically practical. To achieve high throughput, irradiator facilities use large numbers of high-activity radiation sources. As mentioned in Chapter 2, the activity in cobalt-60 sources in panoramic irradiators accounts for over 98 percent of the total activity in all civilian radiation sources in the United States.

STERILIZATION OF MEDICAL DEVICES

The Food and Drug Administration (FDA) requires that the sterilization of invasive medical devices such as hypodermic needles and scalpels must achieve a sterility assurance level of 10^{-6}.[2] The sterility assurance level is the probability or frequency of contaminated products after processing, so a level of 10^{-6} corresponds to a one in a million chance that one live microbe is in the sterilized load. Three standard sterilization processes are employed worldwide by the majority of single-use medical device manufacturers: gamma irradiation, e-beam irradiation, and ethylene oxide (EO) gas diffusion. Some features of these sterilization methods are summarized in Table 6-1. Also shown in the table are features of autoclave (steam) or dry-heat sterilization, which is usually reserved for multiuse medical devices.

X-ray irradiation is not yet used in a major facility, but it is included with gamma irradiation because x-ray generators can meet or exceed the specifications for gamma irradiators listed in the table. The critical differences between these two types of irradiation are discussed in the section of this chapter on x-ray irradiators.

Radiation Processing for Sterilization of Medical Devices

Because gamma radiation penetrates through a product, killing pathogens along its path, yet does not heat the packaging or the product significantly, it can be used to sterilize devices already sealed in heat-sensitive, air-tight plastic packaging. This is a significant benefit for some single-use medical devices and kits, such as those containing hypodermic needles preloaded with a pharmaceutical. Gamma irradiation has proven performance in killing pathogens and is one of the preferred methods, as evidenced by the quantity of product irradiated each year. To achieve a 10^{-6} sterility assurance level requires a dose in the range of 15 to 40 kGy (commonly 25 kGy) at the most shielded point in the package, per ISO Standard 11137.

[2] A sterility assurance level of 10^{-3} is used for many noninvasive medical devices. The doses associated with this level are lower and "kinder" to materials, especially in new drug/device combination products. ISO validation methods (ISO 11137-1, -2, and -3, and VD Max) allow for differing doses and product sterility assurance level depending on bioburden and product use.

TABLE 6-1 Methods of Medical Device Sterilization and Their Features

	Method			
	Autoclave or Dry Heat	Gamma or X-ray Irradiation	Electron Beam Irradiation	Ethylene Oxide (EO)
Processing mode	Batch	Continuous or batch	Continuous	Batch
Post-treatment testing required for product release	None	None	None	Testing required
Part of product sterilized	Mostly surface	Complete volume	Complete volume, but for limited thickness	Surface, with the use of gas-permeable packaging
Material compatibility	Heat-tolerant product and packaging	Most materials are satisfactory Can be incompatible with, e.g., PVC, acetal, polypropylene homopolymer, and polytetrafluoroethylene (PTFE)	Most materials are satisfactory Similar to gamma regarding compatibility, although may have lesser oxidative effects	Nearly all materials are compatible
Residuals	None	None	None	Ethylene chlorohydrin, requires aeration after processing
Best process match according to sterilizers	Reusable devices	Products that are medium to high density, and somewhat heterogeneous	Products that are low density, homogeneous, tolerate high dose rates, and in thin packages	Products that cannot tolerate irradiation, including both single-use and reusable devices

NOTES: In batch mode, a whole batch of product packages undergoes sterilization together. In continuous mode, the product packages are sterilized sequentially. SOURCE: Adapted in part from Sterigenics, Inc. (2007).

There are limitations to the use of gamma irradiation based on dose rate and radiation effects in the device material. For example, thick metal parts on a device can act as shields, resulting in low doses in shadowed locations on the device (this is more of a problem for e beam irradiation). Some plastics discolor or become brittle upon irradiation, although there has been some progress in development of radiation-resistant polymers.[3] But for the many products for which gamma irradiation is effective, manufacturers need only consider other business

[3] Polyimide, liquid crystal polymer, polyether sulphone, polyetheretherketone, polyethylene terephthalate , and other similar plastics can all be made to be relatively radiation resistant, but many other common polymers such as polyoxymethylene, and polypropylene have poor radiation resistance. Radiation-stabilized grades of these latter polymers have been developed to improve their performance under radiation.

factors when choosing a sterilization method. Because medical devices and supplies tend to be bulky, low-density products, manufacturers consider the proximity of the sterilization facility and the timeline for processing along with costs.

The radiation field in an irradiator is virtually constant during the period of irradiation, but the dose delivered within the product depends on the materials to be irradiated, the density, and the thickness of the product. Each product-package combination requires a dose plan and dose map. Some irradiators do not charge customers for this service directly if the customer is contracting for irradiation of the product on a large scale, but will charge for dose mapping for smaller contracts. For customers that sterilize small batches of products or that need very fast turn time from manufacture to delivery, there may be a market for in-house irradiation (rather than contract irradiation) if economic, relatively simple, and appropriately sized irradiators can be developed.

Large-Scale Gamma Irradiation

Facilities that carry out large-scale irradiation using radionuclide radiation sources (gamma sources) rather than x-rays or electron beams have large quantities of radioactive material. A typical commercial panoramic irradiator facility may have 110,000 TBq (about 3 million Ci) of cobalt-60, and some have two times that amount. In a panoramic irradiator, the products to be irradiated pass around high-intensity radiation sources inside a shielded room. While in use, the irradiation room has physical and procedural measures in place to prevent worker access. When the source racks in a wet-storage irradiator are not in use, they are lowered into a pool below the irradiation room. The pool provides shielding and cooling.[4] Figure 6-1a shows an irradiator in which products are passed around a rectangular source rack using hanging tote boxes. Figure 6-1b shows a less common dual cylindrical source rack used in an irradiator that carries products in their shipping pallets on a conveyer system. Modern irradiation facilities are fully automated, so workers need not enter the irradiation chamber to emplace the product. Simply for safety purposes, these facilities are much more robust than ordinary industrial structures and have security controls in place. Additional security measures have been required for these facilities in recent years (see Chapter 3 for a brief discussion of security issues related to panoramic irradiators).

Replacement Technologies

As noted above, sterilization can be carried out by irradiation technologies, heat, or EO diffusion. Each of these options is described below.

E-Beam Irradiation

E-beam irradiators use an accelerator to direct an energetic beam of electrons (usually 5 to 10 MeV) at the product (see Figure 6-2). The beam is scanned across the product in a pattern that ensures that the whole face of the package receives a relatively uniform flux of electrons. The electrons from the beam (the primary electrons) transfer their energy to electrons in the atoms of the product (secondary electrons), which are knocked free and in turn transfer their energy to other electrons in the product. This cascade of electrons delivers its dose throughout the product. Because the beam can be aimed, virtually all of the beam energy can

---

[4] One company, GrayStar, Inc., offers a design in which the source rack remains in a pool at all times and the watertight product totes are lowered into the pool for irradiation.

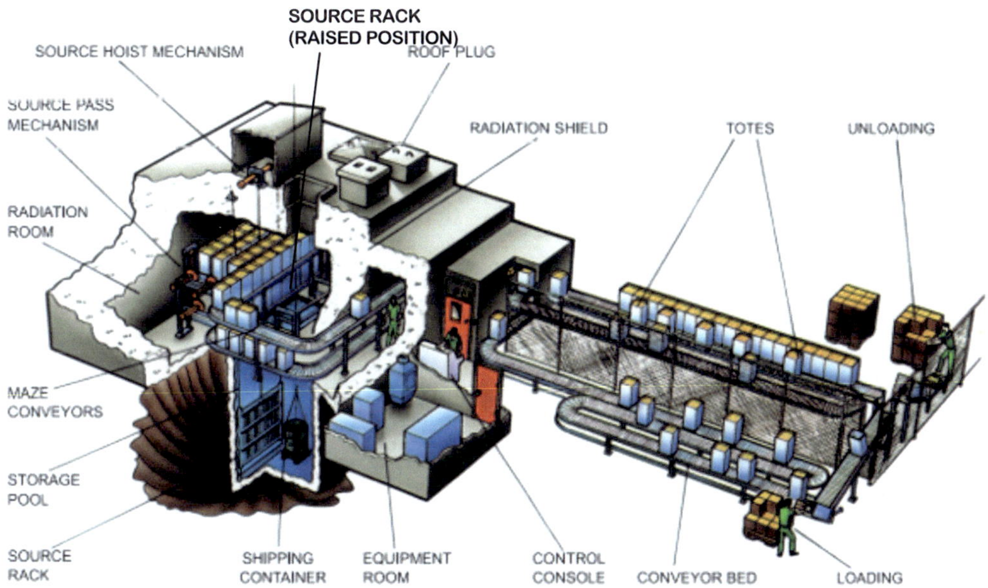

(a)

(b)

FIGURE 6-1 (a) An artist's rendition of an MDS Nordion JS-10000 panoramic irradiator, which uses a panel-type source rack (not necessarily to scale), and (b) a photograph of different source racks for an MDS Nordion Pallet Irradiator. SOURCE: Images courtesy of MDS Nordion (2002).

be directed at the product, although some of the energy exits the product through bremsstrahlung (x-rays from electron collisions) and electrons near the surface that escape.

Electrons transfer energy very efficiently to other electrons because they interact with every electron along their path. The x-rays and gamma rays interact more weakly with fewer electrons. Figure 6-3 shows the relative dose versus depth in material for four different radiations: e-beam at 10 MeV, cobalt-60 with its 1.3-MeV gamma rays, and x-rays at 5 and 7 MeV. The depth of penetration (dose as a function of depth) depends on the density of the material and so the dose-depth relationship is characterized by the product of density (g/cm^3) and distance (cm), yielding units of grams per square centimeter (g/cm^2), rather than actual depth.

A given dose-depth value, say 1 Gy at 1 g/cm^2, implies that lower densities, for example, 0.5 g/cm^3, result in deeper radiation penetration, 1 Gy at (1 g/cm^2)/(0.5 g/cm^3) = 1 Gy at 2 cm. The chart shows that the electron beam delivers its whole dose in a small depth, whereas gamma rays and x-rays spread their doses over a greater depth. The targeted delivery of the electron beam and the ability to deposit nearly all of the energy in a shallow depth enable e-beam irradiators to achieve much higher dose rates than other technologies, which makes it the preferred technology for some applications.

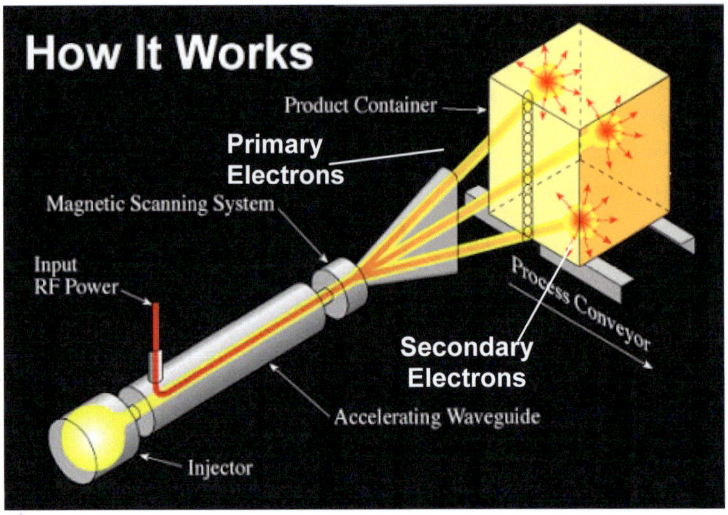

FIGURE 6-2 E-beam irradiation. SOURCE: Image courtesy of GAO (2002).

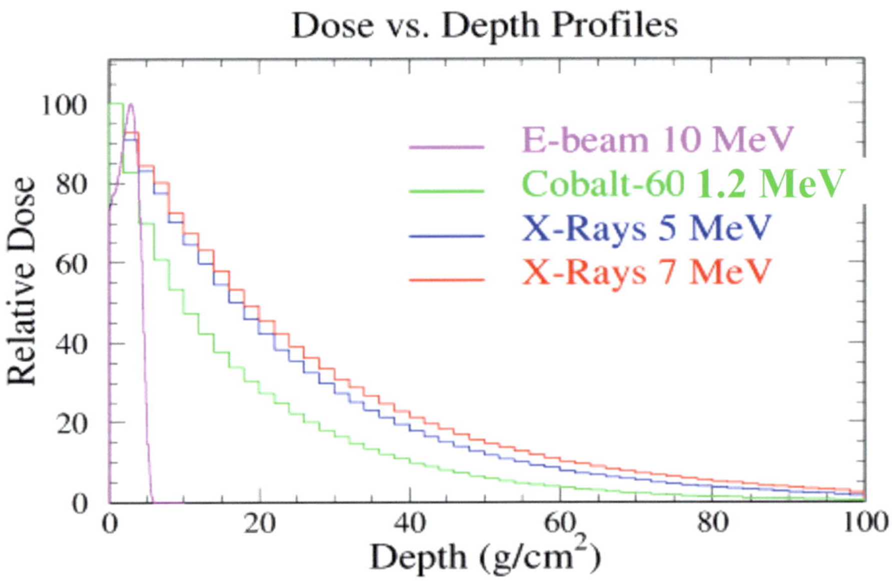

FIGURE 6-3 Relative dose versus depth in material for four different radiations. The highest relative dose for all depth values greater than 5 g/cm^2 is the 7-MeV x-ray. The 5-MeV x-ray is just below that, and the cobalt 60 curve is next. The e-beam curve drops to zero relative dose at about 6 g/cm^2. SOURCE: Image courtesy of Cleland, M. (2006).

As is described in Chapter 4, accelerators must convert electrical power to beam power. The conversion efficiency ranges from 20 to 45 percent, depending on the design of the system and the power output of the irradiator (high efficiency for high-power systems, in general).

X-ray Irradiation

A large-scale x-ray irradiator looks nearly identical to an e-beam irradiator, and can even operate using the same accelerator, but has a target that converts the e-beam to x-rays (see Figure 6-4). The target is a thin layer of high-atomic-number, dense material that can withstand high heat loads; usually tantalum or tungsten is used. The electron beam strikes the target which stops the beam electrons in a short distance. Slowing and stopping the electrons releases radiation called bremsstrahlung. This "braking radiation" is scattered forward in the direction the electrons were traveling, so the accelerator and target generate a fan beam of x-rays with energies up to the energy of the impinging electrons. The x-rays are not the monoenergetic like the gamma rays released when cobalt-60 decays, but the energy of the x-rays can be adjusted. A major advantage of the x-ray irradiators is the ability to use higher energy x-rays: commonly 5 MeV but higher energies are possible.

X-ray irradiation so far has only been used for food irradiation in laboratory and demonstration-scale irradiators and in one moderate-size facility for irradiating packages, described below. Texas A&M University hosts the National Center for Electron Beam Food Research, which does research, training, and contract processing using linacs that deliver e-beam or x-ray irradiation. IBA-Sterigenics constructed a 170-kW facility in Bridgeport, New Jersey, with one 10-MeV e-beam for polymer processing and two x-ray beam lines (one at 5 MV and one at 7 MV) for food irradiation. The 7-MV beam line was constructed at least in part to petition FDA to raise the 5-MeV energy limit for food irradiation, which FDA did. However, the facility won a contract for irradiation of mail for the U.S. Postal Service. The e-beam operation is dedicated entirely to irradiation of flat mail, and the 5-MeV x-ray line is dedicated to irradiation of bulky parcels. The 7-MeV x-ray line is operational but not used. Sterigenics split from IBA, but IBA still has a contract to construct a major x-ray irradiator facility for Sterigenics in Belgium. The companies disagree about whether the new facility will be economically competitive with gamma irradiators. Certainly, higher energy e-beams have better energy conversion efficiency: The conversion efficiency is 8 percent for 5 MeV and about 11.2 percent for 7 MeV. Some supporters of x-ray irradiation have concluded that larger x-ray facilities (several hundred kilowatts) will have economic advantages, and this facility, at around 700 kW, will test that conclusion. The maximum energy used today is 7.5 MeV because of concerns about neutron production and induced radioactivity in the sterilized product.

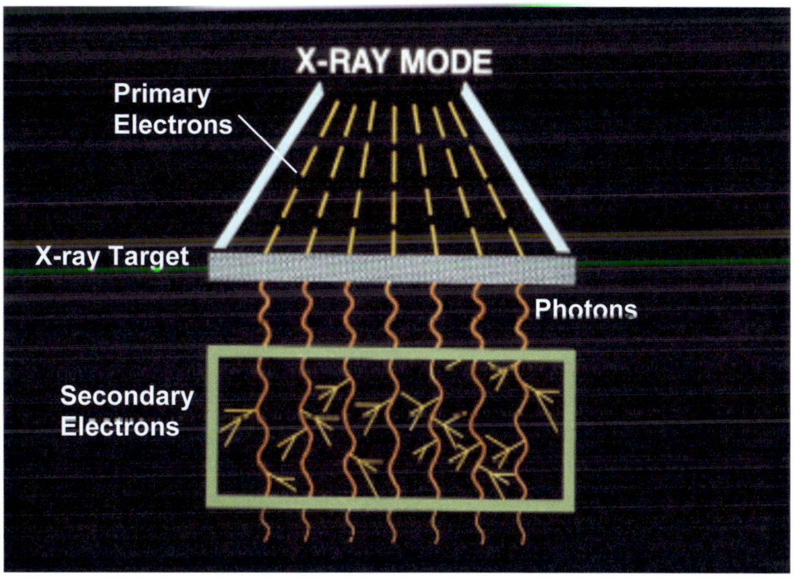

FIGURE 6-4 X-ray irradiation. SOURCE: Image courtesy of Cleland, M. (2006).

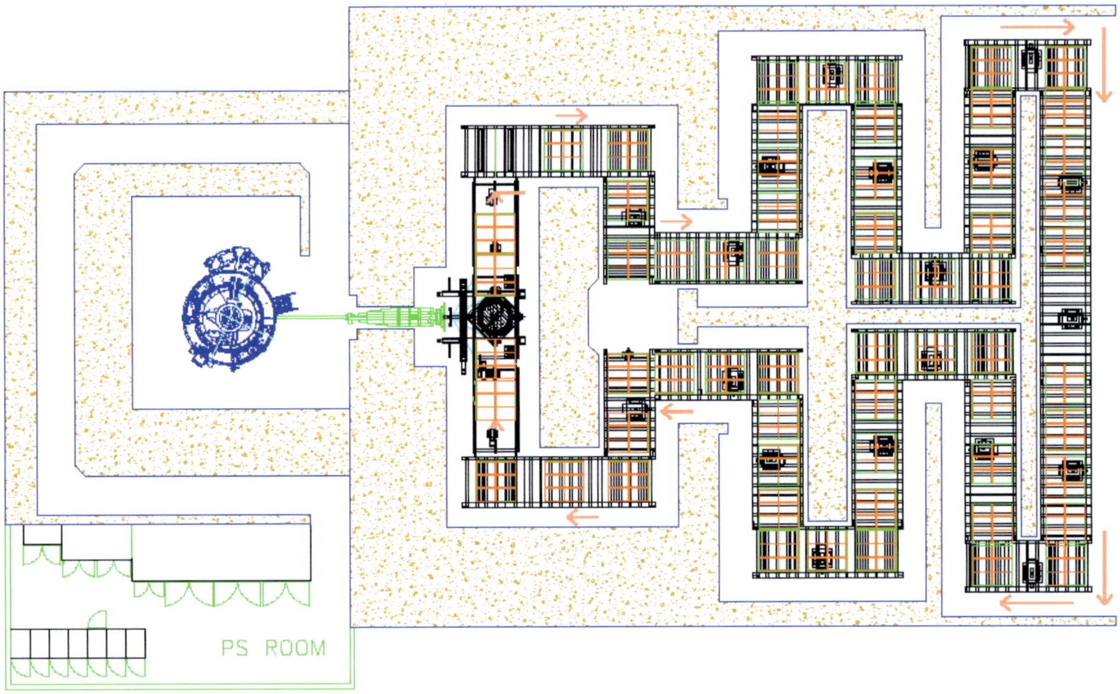

FIGURE 6-5 Conceptual plan view of an x-ray irradiator using the IBA Rhodotron for the electron beam and the Palletron for handling the products. SOURCE: Image courtesy of Cleland, M. (2006).

IBA's concept for an x-ray irradiator facility is illustrated in Figure 6-5, which shows the IBA Rhodotron accelerator as a near-circular device on the left and the products passing through the shielding maze on a conveyor belt, on the right. One pallet or a set of pallets undergoes irradiation at any given time. To even out the dose distribution within a pallet, the design, called a Palletron, rotates the product. Figure 6-6 illustrates the dose as a function of depth in a package irradiated from two sides with cobalt-60. Similar but flatter total dose distributions can be achieved with 5-MV x-ray sources. An alternative facility design has three rotating pallets, one behind another in line with the x-ray beam. This design takes advantage of the fact that a pallet of low-density materials provides only modest shielding of 5-MV x-rays, so much of the x-ray energy can be utilized even in the shadow of another pallet.

Three companies, Mevex Corporation, Precision X-ray, Inc., and RadSource Technologies, Inc., told the committee that they are developing or are willing to develop specialized x-ray irradiation systems to meet the demands of customers that want in-house irradiation to sterilize small batches of products. Titan Scan also offered a small batch e-beam system. Other x-ray tube and compact-accelerator manufacturing companies might also be interested if the market were sufficiently large. Varian, for example, indicated an eagerness to develop x-ray systems tailored to the needs of specific applications if a clear and sizable market were apparent. These devices would most likely be, in essence, self-contained irradiators, but they could replace some contract irradiation if the costs of purchase and operation turn out to be competitive.

An effective system for irradiating a wider variety of products could be imagined as including a combination contract irradiation facility with a gamma irradiator for moderate- and high-density products and for low-dose-rate irradiation; an e-beam irradiator for low-density products and very-high-dose-rate irradiation; and an x-ray irradiator for high- and very-high-density products and high-dose-rate irradiation. Co-located gamma and e-beam irradiation facilities already exist and are operated in the United States. Utilizing the same accelerator that makes the e-beam to make an x-ray line avoids the cost of another accelerator (one still needs the new beam line and x-ray target, the shield, and conveyor), but it is not clear that there is a sufficient market of goods for which x-ray irradiation is sufficiently superior to warrant construction of a separate facility when a gamma irradiator is already available.[5]

Cost Comparison of Gamma and X-ray Irradiation

The clearest comparison of irradiation techniques is between cobalt gamma irradiation and high-energy (7–10 MV) x-ray irradiation. Several cost factors are the same for both: the cost of land, the maze leading into the chamber, the warehouse, and office space. Morrison (1989) notes that the cost of shielding and the conveyor system for cobalt-60 facilities increases with designed hourly throughput because the irradiation chamber must be larger. Accelerators increase throughput by increasing the beam power and conveyor speed, and so the configuration changes little. The factors that more clearly differentiate the cost of cobalt gamma irradiation from x-ray irradiation are listed in Table 6-2. The approximate costs are calculated for irradiators sized to handle roughly 119,000 m^3/yr (4.2 million ft^3/yr) at 25 kGy.

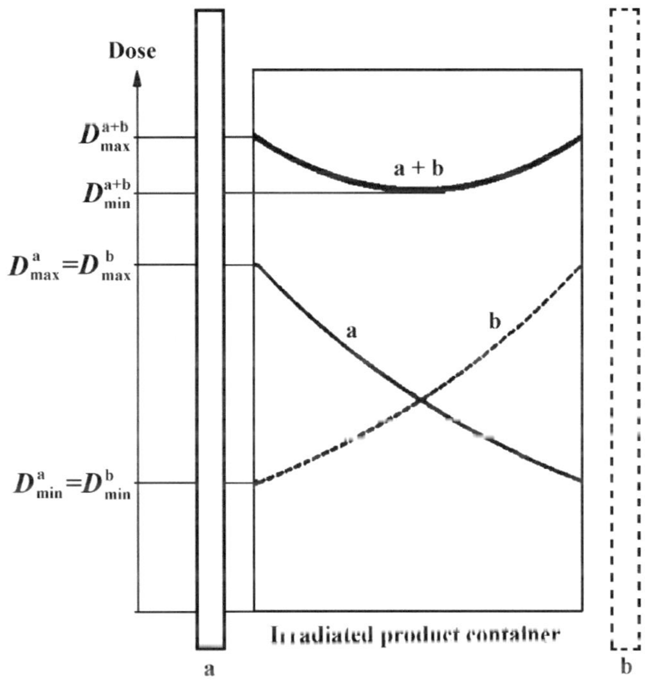

FIGURE 6-6 Depth-dose distribution in a product container irradiated from opposing sides with a cobalt-60 source. Curve a represents the depth-dose distribution when the product is irradiated by a source rack in positiona. Curve b is for the source rack in position b. Curve a + b shows the sum of doses from irradiation on both sides. SOURCE: Image courtesy of IAEA (2004b).

[5] To the committee's knowledge, one existing gamma irradiation facility was converted to e-beam irradiation for research and development at a company's headquarters. It is not at all clear that this option is cost-effective for a production operation because the configuration of the shield and the conveyor system for the products are different for the two irradiators. The conversion would also requires cutting holes in the 2-m-thick reinforced concrete shielding to enable the beam line or the radiofrequency energy to pass through.

TABLE 6-2 Comparison of the Costs of Cobalt Gamma and X-ray Irradiation for Sterilization

Expense Category	Cobalt Gamma Irradiation	X-Ray Irradiation
Initial investment	Initial loading of cobalt-60 130,000 TBq = $7M (assuming $54/TBq [$2/Ci])	Electron accelerator and targets 7 MeV, 78.4 kW (x-ray power) = $7M
	Facility costs including shield, conveyor system, hoists, and wet-storage pool = Approximately $5M	Cost of shield and conveyor system = approximately $5M
Operating costs	Cobalt-60 replenishment 12.3%/y = $0.86M/yr Regulatory costs = unknown but higher for gamma irradiator Security = unknown but higher for gamma irradiator	Electricity for the accelerator 1,400 kWe = 11,040 MWh /yr[a] or about $660,000/yr[b] Maintenance and operations = perhaps $0.25M/yr more than gamma irradiator[c]
Decommissioning and disposal costs	Final return shipment of cobalt-60 = $0.25M	Disposal of spent targets = $0.02M

NOTES: Categories with no significant cost difference, such as land, office space, laboratory costs, and tear-down of the facility at decommissioning, are not listed.

[a] Assuming the equipment is 50% efficient for the e-beam and operates 90% of the time

[b] $59.7 per MWh nationwide average for industrial price of electricity based on data from EIA (2006).

[c] Three highly skilled maintenance technicians for the x-ray system versus one for the gamma irradiator.

SOURCE: Adapted from Cleland, M. (2006); Smith, M. (2006); and Morrison, R. (1989).

These rough calculations suggest that x-ray irradiation is economically viable if the assumptions about performance and costs hold true. Because there is no experience yet with a large-scale x-ray irradiator, the committee cannot state these assumptions with great confidence, and only offers them as the data it has available.

Steam or Dry Heat

Autoclaves and dry-heat ovens are routinely used in hospitals to sterilize reusable medical devices, but medical devices are increasingly being provided as single-use devices. Autoclaves are essentially pressure cookers used to sterilize devices and equipment. Heating water in a sealed enclosure increases its boiling point as the pressure increases. This enables the water to reach temperatures well above 100°C. Dry-heat ovens operate at higher temperatures, but their heat transfer properties are less efficient, so the sterilization takes longer unless operated in a convection mode by blowing air on the products. Table 6-3 shows sterilizer temperatures, pressures, and times recommended in an article in the *Journal of the American Dental Association*. Steam is a surface sterilizer, unless the entire device remains for a sufficient time at a temperature that kills pathogens. Chemical vapor sterilizers called chemiclaves are also used.

Other emerging technologies have been applied to small-batch sterilization devices for use in hospitals, not for contract irradiation. For example, hydrogen peroxide plasma (Rutala and Weber, 2001) and supercritical carbon dioxide (White et al., 2006) are being marketed as replacements for compact EO sterilizers.

TABLE 6-3 Autoclave and Dry-Oven Temperature and Time Pressures for Sterilization with EO Included

Sterilizer	Temperature (°C [°F])	Pressure (kPa [psi])	Time
Steam autoclave	121 [250]	100 [15]	15 min
unwrapped items	132 [270]	200 [30]	3 min
lightly wrapped items	132 [270]	200 [30]	8 min
heavily wrapped items	132 [270]	200 [30]	10 min
Dry heat, wrapped	170 [340]		60 min
	160 [340]		120 min
	150 [300F)		150 min
	140 [285F)		180 min
	121 [250F)		12 h
Dry heat (rapid flow) unwrapped items	190 [375F)		6 min
packaged items	190 [375]		12 min
Chemical vapor	132 [270]	140-280 [20-40]	20 min
EO	40–60 [100–140]	6.7-50 [1–7.5]	Varies, 4–10 hr

SOURCE: Courtesy of Journal of the American Dental Association (1991).

EO Sterilization

More than half of all sterile medical devices sold are sterilized using EO (J. Hadley, Ethylene Oxide Sterilization Association, Inc., personal communication to F. San Martini, 2007; Dever et al., 1994). EO is used for most of the medical products that are incompatible with radiation exposure (Hadley, personal communication, 2007). EO can be used for most current hospital surgical kits, catheters, IV tubing, endotracheal tubing, angiographic balloons, heart kits, cranial and orthopedic implants, pacemakers, and implantable deflbrillators. In addition, more than one-third of all reusable devices are currently sterilized with EO (Hadley, personal communication, 2007). Some medical products are not suitable for EO sterilization (Hadley, personal communication, 2007), such as vacuum-pressure-sensitive products; nonvented, sealed products that do not allow for gas diffusion or penetration of EO gas; medical products that retain absorbed EO; products with extremely high densities or challenging physical configurations that would limit the permeation of EO; products with active pharmaceutical ingredients not validated for the effects of EO; and some orthopedic implants where radiation is needed to increase product strength characteristics, in addition to sterilization.

EO sterilization typically comprises three stages: preconditioning, sterilization, and aeration. In a conventional sterilization process, these steps are done separately. In some cases, however, the three steps are completed together within the sterilization chamber. The four key parameters affecting EO sterilization efficiency are EO concentration, temperature, relative humidity and exposure time. Because of differences in packaging, load density, and other factors, each product type requires a unique treatment cycle.

Products to be processed are placed on pallets and then enter the preconditioning phase, which helps ensure EO penetration. Preconditioning consists of exposing products to elevated temperatures (typically between 40°C and 60°C) and relative humidity levels (typically between 45 and 75 percent) for up to one day.

Sterilization takes place in a stainless steel sterilization chamber. Commercial sterilization chambers vary considerably in size, ranging in capacity from 1 to 30 pallets (Hadley, personal communication, 2007). Typically, the sterilization process is performed under negative pressure conditions (below atmospheric pressure). Once vacuum has been established, EO is pumped into the sterilization chamber.[6] This period, during which products are exposed to high EO concentrations, is termed the *dwell stage*. The EO concentration and cycle times vary greatly depending on the product, cycle conditions, and whether a conventional EO sterilization process is used or the three sterilization stages are performed together in the sterilization chamber. The EO concentrations are highest during the dwell stage and range from about 300 to 1,200 mg/l, with the average at about 650 mg/l (Hadley, personal communication, 2007).[7]

If the three sterilization stages are performed separately, the chamber time typically ranges from 8 to 12 hours. If the three steps are performed within the chamber, the total chamber time is about 11 to 36 hours (Hadley, personal communication, 2007). After the dwell stage, EO from the chamber is exhausted to air pollution control equipment with successive gas (typically nitrogen) washes. After completion of post-sterilization flushing, the product is transferred to the aeration stage, which removes residual EO. Finally, the products must be tested to verify sterility.

EO is extremely flammable, and gas/air mixtures are explosive. The flammability limits in air are 3 percent (30,000 part per million by volume [ppmv]) to 100 percent (Lewis, 2003); pure EO can be ignited in the absence of air. According to the National Institute for Occupational Safety and Health (NIOSH), once ignited, it can flash back to the fuel source with velocities of 1,800 to 2,400 m/s (NIOSH, 2000). The gas is colorless, heavier than air, and may travel along the ground; distant ignition is possible (IPCS, 2001). EO can be detected by odor only when it has already reached the dangerous concentration of 260 ppm (NIOSH, 2000). EO is reactive with strong acids; alkalis and oxidizers; chlorides of iron, aluminum, or tin; and oxides of iron and aluminum (Lewis, 2003).

EO is a "known" or "probable" human carcinogen, depending on the classifying body.[8] Repeated or prolonged inhalation exposure may cause asthma; it may have effects on the nervous system, liver, and kidneys, or cause cataracts; and it may cause heritable genetic damage to human germ cells. There are reports of EO-induced anaphylaxis from sterilized

---

[6] Most commercial sterilization facilities currently use pure EO (Hadley, personal communication, 2007). Previously, mixtures of EO and chlorofluorocarbons (CFCs) were used to reduce flammability and the risk of explosions. EO–CFC mixtures were phased out following the Montreal Protocol.

[7] Assuming sterilization occurs at approximately 0.67 atm and room temperature (298.15 K), the range of EO concentration during the dwell stage is approximately 250,000 to 999,700 ppmv, with an average of approximately 540,000 ppmv (i.e., 25–99.97 percent, with an average of 54 percent).

[8] The National Toxicology Program recently upgraded EO to a known human carcinogen. In 1985, the U.S. Environmental Protection Agency classified it as a Group B1 (probable) carcinogen; a new draft evaluation of the carcinogenicity of EO was being evaluated at the time of this writing. The International Agency for Research on Cancer (IARC) classified it as a Group 1 carcinogen. NIOSH (2004) found that persons exposed to very high levels of EO may be at an increased risk of developing blood cancers among men and breast cancers among women.

membranes used in hemodialysis (Ebo et al., 2006). Both chronic and acute exposures may cause miscarriages.

Because of its environmental, safety, and occupational hazards, EO is regulated by federal and state agencies. Oxidizing emission control devices are generally used to remove EO from low-concentration emission streams; acidified wet scrubber systems are typically used when emissions contain high EO concentrations.

The Occupational Safety and Health Administration (OSHA) permissible exposure limit for EO is 1 ppmv and the NIOSH recommended exposure limit for EO is 0.1 ppmv, both as an 8-hour time-weighted average, and a short-term exposure limit of 5 ppmv, time-weighted over 15 minutes (29 CFR § 1910.1047). NIOSH has determined that 800 ppmv is the EO concentration that is immediately dangerous to life and health. OSHA also requires:

- monitoring employees to determine actual exposure to EO during work shift,
- restricting access to EO areas to authorized personnel, and
- implementing a system to provide emergency warning in the event of a release.

In addition to the requirement to meet the sterility assurance level described above, FDA regulations specify permissible residual concentrations of EO on sterilized medical products.

Safety and Security and EO

To educate users of EO, five chemical companies produced and make freely available (http://www.ethyleneoxide.com) a guide that summarizes essential information for safely handling EO (Buckles and Chipman, 1999). The summary of incidents involving EO given in Section 5 of the guide illustrates EO's hazards. These range from a railcar[9] explosion that caused major damage over a 300-m radius and broke windows up to 5 km away; and an incident in which 0.27 kg of EO decomposed in a pump, causing the upper part of the pump and its motor (weighing approximately 450 kg) to break free of the 12 steel bolts that held them in place and shoot over 18 m into the air (Buckles and Chipman, 1999).

More recent examples of accidents at EO sterilization facilities confirm that process safety concerns remain. An explosion at an EO sterilization facility in California in 2004 injured four workers and severely damaged the facility. The explosion sheared the hinges off both of the 1,800-kg (4,000-lb) sterilization chamber doors and propelled them outward. One door came to rest approximately 25 m from the chamber after striking and fracturing the south wall of the building, while the other came to rest approximately 5 m away, after colliding with and damaging a steel column (U.S. Chemical Safety and Hazard Investigation Board, 2006).

A similar explosion occurred at a sterilization facility in 1997. The explosion occurred during a test of a newly installed oxidizing emission control device that replaced an acidified wet scrubber system. The explosion blew off the sterilizer door and moved the 22,700-kg (50,000-lb) sterilization chamber off its foundation. About 7 to 9 kg (15 to 20 lb) of EO is believed to have been in the sterilizer at the time of the explosion (NIOSH, 2004).

---

[9] This incident involved an EO shipment from a European EO producer to a customer. In North America, all EO sterilization companies receive their EO supply in gas drums that are shipped via truck from ARC Specialty Products. These drums comply with current Department of Transportation requirements. They are double-walled stainless steel drums that provide protection for all valve openings. The drums are regularly inspected, pressure tested, and drop tested (Hadley, personal communication with F. San Martini, 2007).

EO Summary

EO gas sterilization has been used for more than seven decades.[10] Less than 1 percent of all EO produced in the United States is used as an industrial sterilant or fumigant (LaMontagne and Kelsey, 1998). Therefore, concerns over its hazards should be viewed in context. It is a technology that has been used for many years to sterilize medical products effectively. There have been no accidents or incidents involving mass casualties due to the use of EO. However, there are substantial health and safety concerns surrounding the use of EO and encouraging substitution to EO just replaces one kind of risk with another.

FOOD IRRADIATION

Foods such as spices, fresh fruit, vegetables, and grains can be irradiated to slow the ripening process, prevent sprouting, extend shelf-life, and kill bacteria, parasites, and mold. Meat and poultry can be irradiated to 4.5 kGy for similar purposes. Table 6-4, taken from Deeley (2001), lists the typical doses required for different food irradiation applications, and Table 6-5 lists the approved uses of radiation for treatment of food. Food irradiation, which can use the same kinds of equipment as sterilizer irradiators (cobalt-60, e-beam, or x-ray), is not common in the United States, but it may increase in the future.

TABLE 6-4 Typical Radiation Doses for Various Food Applications

Application	Food Product	Typical Dose (kGy)
Reduction or elimination of microbial populations in dry food ingredients	Spices Starch Enzyme preparations	3–10
Pasteurization	Meat Poultry Shellfish Frogs' legs Herbs/spices	2–7
Extend shelf–life	Fruits Vegetables Meat, poultry Fish	0.5–5
Parasite disinfection	Meat Pork Fish	0.1–3
Insect de-infestation	Grain Flour Dried fruits	0.2–0.8
Inhibition of sprouting	Onions Garlic Potatoes	0.03–0.14

NOTE: The necessary radiation dose depends on the application and the bacteria being treated. Moisture reduces the necessary dose. Absorbed dose is measured in Gray (Gy). 1 Gy is 1 J of energy absorbed per kilogram of food irradiated. SOURCE: Adapted from Deeley (2001).

[10] Application for a patent for sterilization using EO was made as early as 1933 by Gross and Dixon (1937).

TABLE 6-5 Approved Uses of Radiation for Treatment of Food in the United States

Year Approved	Food	Dose (kGy)	Purpose
1963	Wheat flour	0.2–0.5	Control of mold
1964	White potatoes	0.05–0.15	Inhibit sprouting
1986	Pork	0.3–1.0	Kill Trichina parasites
1986	Fruit and vegetables	1.0	Control Insects, increase shelf life
1986	Herbs and spices	30	Sterilization
1990, FDA	Poultry	3	Reduce bacterial pathogens
1992, USDA	Poultry	1.5–3.0	Reduce bacterial pathogens
1997, FDA	Meat	4.5	Reduce bacterial pathogens
1999, USDA (pending)	Meat	4.5	Reduce bacterial pathogens

SOURCE: Courtesy of CDC (2005).

The largest food irradiator in the United States, SureBeam, declared bankruptcy in 2004, and its facilities, which handled large quantities of ground beef, shut down. Since then, a company called BeamOne, LLC, has operated the former SureBeam facilities in San Diego, California, Denver, Colorado, and Lima, Ohio for sterilization of medical products and for polymer processing. Sadex purchased SureBeam's e-beam irradiator in Sioux City, Iowa.

Currently, two commercial facilities in the United States routinely irradiate food: one is in Florida (Food Technology Service, Inc., a gamma irradiator) and the other is in Hawaii (Hawaii Pride; an e-beam facility rated at 15 kW). In August 2007, Hawaii Pride received a license to build another facility in Hawaii, this one using cobalt-60. In addition, the U.S. Department of Agriculture (USDA) has one gamma irradiation facility dedicated to food irradiation research, Iowa State University operates the Linear Accelerator Facility for food irradiation, and Texas A&M University operates the National Center for Electron Beam Food Research, a semi-commercial, semi-research facility for e-beam and x-ray food irradiation. A few other irradiation facilities can be contracted to irradiate foods, but do not do so routinely (aside from spices) and do not have a refrigerated storage warehouse for receiving products.

As noted earlier, the same kinds of equipment can be used for irradiation of food as for sterilization irradiation. E-beam irradiators compete with gamma irradiators in this market today. X-ray irradiators face the same economic uncertainties for food irradiation as for sterilization applications. While fumigation and chlorine rinses are possible for produce, there is no direct replacement for irradiation of ground beef, which is currently the biggest market for U.S. food irradiators.

MATERIALS PROCESSING USING PANORAMIC IRRADIATORS

Although the majority of irradiators are devoted to medical device sterilization, contract irradiators (gamma and e-beam) are also used in a number of materials processing applications. These range from the irradiation of PTFE (Teflon®) to create PTFE micropowders, useful in inks and lubricants, to the treatment of polymer strips to enhance their lubricity). These applications depend on ionizing irradiation to break chemical bonds in polymers and to decrease their molecular weight. Because the density of polymers is low, this can be accomplished by ionizing radiation produced by x-ray, e-beam, or gamma rays.

Doses cited for materials processing range from 200 kGy for cross-linking of polymers to well over 500 kGy for degradation of PTFE. High dose rates are desirable for such high doses, and so e-beam irradiation is the preferred approach for some of these applications. Indeed, it is the primary application of e-beam irradiators and, for very thin targets, of high-dose-rate kilovoltage x-ray sources. Many of these materials processing applications appear to be proprietary processes and it is not clear to the committee how extensively gamma irradiation facilities are used for these purposes. Chemical cross-linking is the predominant technique used by polymer manufacturers, and so those using irradiation have specifically sought out the features peculiar to irradiation.

INSECT STERILIZATION

USDA operates about a half-dozen irradiators for sterilization of male insects, called the sterile insect technique. This practice is carried out extensively outside the United States (see, e.g., Enserink, 2007a,b), which constitutes a larger market for the irradiation devices. USDA facilities breed the pests (e.g., the Caribbean fruit fly) in isolated facilities and sterilize live specimens by irradiation with a dose of 100–150 Gy, then load them into transport containers and release them in areas of potential infestation to compete with the infesting populations for breeding. This suppresses the reproduction rate and inhibits infestation. The gamma irradiators USDA uses are decades old and are loaded only with their original cesium-137 sources. As these sources age, the time needed for irradiation lengthens and eventually the devices will need to be replaced. Some irradiation devices utilizing x-rays for sterile insect release have been sold outside the United States (Kirk, 2006). As noted above, too, foods may be irradiated to reduce the threat from pests, such as fruit flies.

FINDINGS

For some applications, alternative technologies to gamma irradiators are already preferred and in use. The primary reason to consider encouraging the sterilization of medical products with EO rather than by irradiation is a reduction of security risk. The risks of EO should be judged relative to those associated with the use of radiation sources in panoramic irradiators. EO poses no area-denial radiological dispersal device risk, but the accidents and potential security risks, health risks associated with exposure, and pressures to encourage EO users to switch *away* from EO because of its toxicity indicate that encouraging a shift from irradiation to EO sterilization may not be desirable.

A direct replacement for gamma irradiators is available and technologically feasible in the form of x-ray irradiators. Whether these replacements are economically competitive with gamma irradiators is not clear. The committee's incomplete calculation shows that the costs could be comparable, but the actual cost differences depend on variable factors such as the cost of electricity and the reliability and throughput of the equipment in the x-ray facility.

CHAPTER 7

RADIOTHERAPY

SUMMARY

Radiotherapy uses ionizing radiation directed at a human or animal body to treat many serious diseases, most notably cancer. High-activity radionuclide sources can be used to create clinical ionizing radiation beams in the form of high-energy gamma rays in teletherapy machines used for external-beam radiotherapy as well as in the Gamma Knife® used for stereotactic radiosurgery. Only four known radionuclides possess characteristics that make them candidates for use in external beam radiotherapy: cesium-137, cobalt-60, europium-152, and radium-226. Europium-152 has not been developed yet for clinical use, and the use of cesium-137 and radium-226 was discontinued for practical reasons and because of safety concerns. Cobalt-60 is currently used in external beam radiotherapy devices found mostly in developing countries. This is because the linear accelerator (linac) is considered a better and more versatile radiotherapy tool, and has largely supplanted cobalt-60 teletherapy devices in the United States and other developed countries. Currently in the United States there are several thousand radiotherapy devices in over 2,400 institutions and clinics (Ballas et al., 2006). Fewer than 250 cobalt-60 teletherapy devices are licensed in the United States and most of those are thought to be in storage for decay, in use for other purposes (such as fixed radiography), or in use for teaching.

Radiosurgery is an irradiation technique that uses radiation beams from many directions to treat lesions. It is now used mainly to treat brain lesions with the Gamma Knife®, a device with approximately 200 cobalt-60 sources. Radiosurgery can also be practiced with isocentric linacs as well as with a miniature linac mounted on a robotic arm, but many neurosurgeons and some radiation oncologists believe that Gamma Knife® radiosurgery is superior to linac-based radiosurgery. The cost of radiosurgery with a dedicated linac is similar to that with a Gamma Knife®, while the cost of radiosurgery with a standard linac modified for radiosurgery is much lower. In addition, linac-based radiosurgery is more versatile for development of stereotactic techniques in the treatment of lesions in organs other than the brain, and on a fractionated (rather than a single-treatment) basis.

Development of new treatment technologies and shifts in practices in radiotherapy are driven by innovation and decisions about delivery of health care, at both the macrolevel (e.g., insurance reimbursements) and at the personal level (i.e., doctor-patient treatment planning). Medical technology researchers already seek out opportunities to improve nonradionuclide-based radiotherapy and radiosurgery. Because linacs have already largely replaced radionuclide-based teletherapy devices, the committee concludes that teletherapy services, while important, are lesser and declining concerns with respect to radiation source security. It is also possible that the recent rapid growth in Gamma Knife® installations in the United States and the developed world will soon subside, given that a viable alternative is offered with linac-based radiosurgery.

INTRODUCTION

Radiotherapy is defined as treatment of disease with ionizing radiation. The diseases treated by radiotherapy are mainly malignant; however, on a smaller scale, radiotherapy is also used in treatment of certain benign conditions. Although the mortality rate for all cancer sites combined has been decreasing, age-adjusted incidence rates for all cancers combined have been stable over the most recent periods of analysis (U.S. Cancer Statistics Working Group, 2006), and the incidence of some cancers is increasing. In developed countries, the increasing life expectancy (around 75 years for boys and 80 years for girls born in 2004 in the United States; NCHS, 2006) combined with low birth rate is causing significant aging of the population and making cancer the second most common cause of death in the United States after heart disease. The American Cancer Society recently estimated that about 1,445,000 new cases of invasive cancer will be diagnosed in Americans in 2007 (ACS, 2007), among a population of about 302 million (about 1 in 210 people), according to the U.S. POPClock Projection of the U.S. Bureau of the Census (U.S. Census Bureau, 2007). Thus, in the developed world, cancer is a major health problem, and treatment of cancer is of major importance to modern societies. The cancer rate is lower in developing countries because people in those countries generally die of causes other than cancer at a younger age. Nevertheless, it is estimated that the incidence of cancer worldwide stands at some 10 million new cancer patients per year.

Radiotherapy is one of three important cancer treatment modalities; the other two are surgery and chemotherapy. In a modern health care system over 50 percent of cancer patients receive radiotherapy either as primary treatment or in conjunction with surgery and chemotherapy.

There are two main types of radiotherapy: *external beam radiotherapy* and *brachytherapy* (also referred to as curietherapy or endocurie therapy). In a typical radiotherapy department about 80 percent of treatments are done with external beam radiotherapy, the rest with brachytherapy. External beam radiotherapy is delivered with the radiation source at a certain distance from the patient, whereas brachytherapy is carried out with radiation sources placed directly into the patient, either into a body cavity (intracavitary brachytherapy) or surgically into a body organ (interstitial brachytherapy).

During the past 100 years of radiotherapy many types of equipment have been designed and used for external beam radiotherapy. Modern external beam radiotherapy is delivered with:

- *x-ray machines* producing kilovoltage x-ray beams, typically used for treatment of superficial (skin) lesions;
- *teletherapy machines* producing megavoltage gamma-ray beams, typically used for treatment of deep-seated lesions;
- *linacs* producing megavoltage x-ray beams, typically used for treatment of deep-seated lesions and megavoltage electron beams typically used for treatment of superficial (skin) lesions.

Brachytherapy sources are most commonly gamma emitters manufactured in the form of a sealed source encapsulated into a special container to prevent leakage of the radioactive material. Because the activity of brachytherapy sources is relatively low (on the order of 0.37 TBq [10 Ci] or less), they are Category 3 sources and as such are outside of this study's scope. Teletherapy external beam sources, on the other hand, are of very high activity (on the order of 370 TBq [10,000 Ci]), warranting Category 1 classification. Linacs are of interest as a viable, safe, and practical alternative to teletherapy machines.

The International Atomic Energy Agency (IAEA) in Vienna, Austria, has recently published a textbook, *Radiation Oncology Physics: A Handbook for Teachers and Students*, which covers in detail the technical and safety issues related to modern external beam radiotherapy and brachytherapy (IAEA, 2005c). Many other textbooks have been published on the technical and clinical aspects of radiotherapy (e.g., Hendee et al., 2004; Khan, 2003; Williams and Thwaites, 2000; Johns and Cunningham, 1984)

STANDARD EXTERNAL BEAM RADIOTHERAPY

Since the inception of radiotherapy soon after the discovery of x-rays by Roentgen in 1895, the technology of radiation production has been aimed toward ever higher photon energies and intensities and more recently toward computerization and intensity-modulated radiation therapy (IMRT). During the first 50 years of radiotherapy, the technological progress was relatively slow and mainly based on x-ray tubes, Van de Graaff generators, and betatrons. All of these units were heavy and bulky and not suitable for isocentric mounting.

The first truly practical megavoltage therapy machine was the cobalt-60 teletherapy machine developed by Canadian physicist Harold E. Johns in the early 1950s (Johns et al., 1951). The invention of cobalt-60 teletherapy provided a tremendous boost in the quest for higher photon energies in a compact unit and placed the cobalt-60 unit into the forefront of radiotherapy for a number of years, mainly because it incorporated a radioactive source that is characterized by features extremely useful for radiotherapy.

Standard e-beam cobalt-60 radiotherapy is carried out with teletherapy machines that are loaded with high-activity (Category 1) gamma-ray sources. The source is most often mounted isocentrically, allowing the beam to rotate about the patient at a fixed source-axis distance (SAD). Modern teletherapy machines have a SAD of 80 cm or 100 cm. Figure 7-1 shows a photograph of a typical modern teletherapy machine using the cobalt-60 radionuclide as the source of radiation. The main components of a teletherapy machine are:

- radioactive sealed source;
- source housing, including shielding, beam collimator, and source movement mechanism;
- gantry and stand in isocentric machines or a housing support assembly in stand-alone machines;
- patient support assembly (treatment couch);
- machine operating control console.

Gamma radiation is obtained from specially designed and built sources that contain a suitable, artificially produced, radioactive material. The parent source material undergoes a beta-minus decay resulting in excited daughter nuclei that attain their ground state through emission of gamma rays (gamma decay).

The important characteristics of a radionuclide used for external beam radiotherapy are:

- high gamma-ray energy (of the order of 1 MeV) for better penetration into tissue;
- high specific activity (of the order of 100 Ci/g) to achieve adequate dose rate with a relatively small diameter source (source size affects the uniformity of the therapy beam);
- relatively long half-life (of the order of several years) to avoid the need for frequent and costly source replacements.

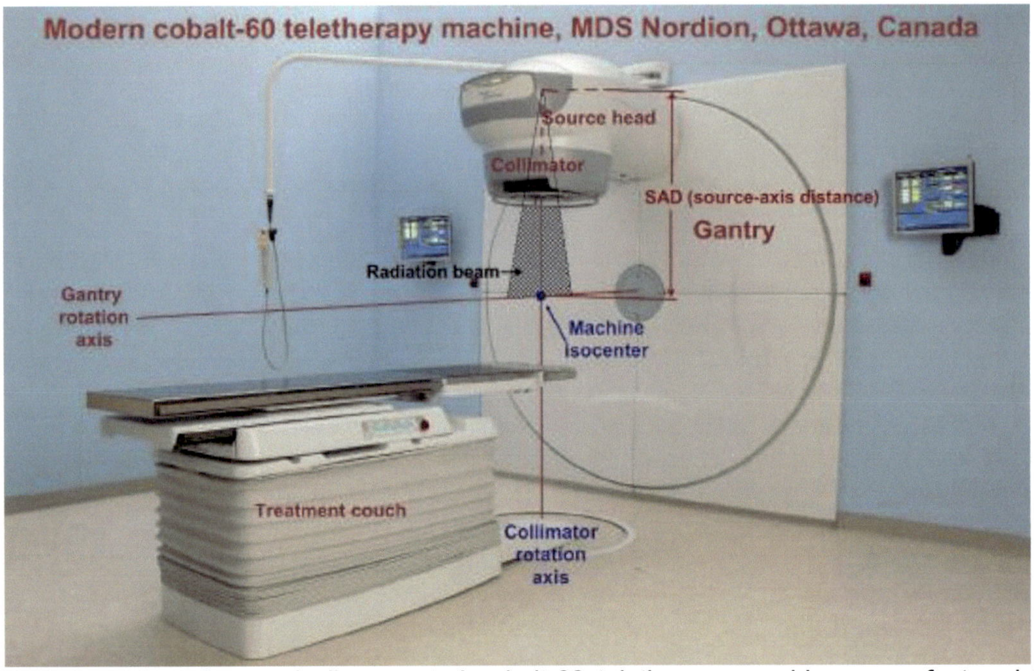

FIGURE 7-1 A modern, isocentrically mounted cobalt-60 teletherapy machine, manufactured by MDS Nordion, Ottawa, Ontario. SOURCE: Image courtesy of MDS Nordion.

The basic physical properties of the two gamma-emitting radionuclides (cobalt-60 and cesium-137) that have proven useful in external beam teletherapy and a potential source for teletherapy (europium-152) are listed in Table 7-1. Of the three radionuclides, cobalt-60 is the most widely used, because it offers the most practical approach to external beam radiotherapy, considering the energy of emitted photons, half-life, specific activity, means of production, and safety. Cobalt-60 and europium-152 come in metallic form encapsulated in a special source container, while cesium-137 is used in the form of cesium-137 chloride encapsulated in a special source container.

The use of cesium-137 for external beam radiotherapy was discontinued during the 1980s despite its attractive half-life. There are two reasons for this: (1) security of cesium sources is a major concern and, in addition, (2) cesium-137 has a relatively low specific activity. The low specific activity implies a relatively low source output and large source diameter; this effectively precludes a teletherapy machine SAD larger than 50 cm, yet in modern radiotherapy a minimum SAD of at least 80 cm is the accepted norm. Note that cesium-137 is still used in blood irradiators and research irradiators because in these machines the distance between the source and the irradiated object is relatively short, of the order of only 10 to 25 cm.

A cobalt-60 teletherapy source is typically a cylindrical stainless-steel capsule containing many hundreds of tiny, high-activity cobalt metal pellets and sealed by welding. A double-welded seal is used to prevent any leakage of the radioactive material from the source container. The typical diameter of the cylindrical teletherapy source is between 1 and 2 cm, the height of the cylinder is about 2.5 cm. For a given activity, a smaller source diameter yields a smaller physical penumbra,[1] making for a sharper beam edge; but higher specific activity sources are more expensive. Often a diameter of 1.5 cm is chosen as a compromise between the cost and penumbra size. The cost of a cobalt-60 teletherapy source is on the order of $200/TBq ($7.5/Ci).

[1] The penumbra is the spread of the beam beyond the idealized beam shape.

TABLE 7-1 Physical Properties of Radionuclides Used In External Beam Radiotherapy

	Radioactive Source		
Property	**Cobalt-60**	**Cesium-137**	**Europium-152**[a]
Half-life (yr)	5.3	30	13.4
Specific activity (Ci/g)	1,100[b] (300[c])	88[b] (30[c])	180[b] (~150[c])
Photon energies (MeV)	1.17 and 1.33	0.662	0.6 and 1.4
Specific gamma rate constant[d] $\Gamma\ [\mathrm{R \cdot m^2 / (Ci \cdot hr)}]$	1.31	0.33	1.06
Specific air-kerma rate constant[e] $\Gamma_{AKR}\ [\mu\mathrm{Gy \cdot m^2 / (GBq \cdot hr)}]$	309	78	250
Half value layer HVL (cm Pb)[f]	1.1	0.5	1.1
Means of production	^{59}Co + *n* in reactor	Fission by-product	^{151}Eu + *n* in reactor

[a] Europium-152 has not yet been developed for clinical use.
[b] Theoretical specific activity: $a = (N_A \ln 2)/(t_{1/2}A)$.
[c] Practical specific activity is smaller than the theoretical specific activity because the source is not carrier-free; i.e., the source contains a stable nuclide in addition to a radioactive nuclide (example.g., cobalt-59 mixed with cobalt-60).
[d] Exposure rate achievable for a given activity of the source.
[e] Dose rate in air for a given activity of the source.
[f] Thickness of shielding (centimeters of lead) required to reduce the radiation by a factor of 2.
SOURCE: Provided by the committee.

To facilitate interchange of sources from one teletherapy machine to another and from one radionuclide production facility to another, standard source capsules have been developed for use around the world. Teletherapy sources are usually replaced within one half-life of installation; however, financial considerations may result in longer source use. Teletherapy machines are designed to enable on-site replacement of sources by trained technicians (the machines are not returned to the manufacturer for resourcing).

The housing for the source in a teletherapy machine is called the source head. It consists of a steel shell with lead for shielding and a mechanism for bringing the source in front of the collimator opening to produce the clinical gamma-ray beam. Two different methods are in use for moving the teletherapy source from the BEAM OFF into the BEAM ON position and back: (1) a source in a sliding drawer and (2) a source in a rotating cylinder. Both methods incorporate a safety feature that terminates the beam (i.e., moves the source into the OFF position) automatically in case of power failure or emergency. When the source is in the BEAM OFF position, a light appears in the BEAM ON position above the collimator opening, allowing an optical visualization of the radiation field, as defined by the machine collimators and any special shielding blocks.

Typical teletherapy source activities are on the order of 185 to 370 TBq (5,000 to 10,000 Ci), and provide a typical dose rate on the order of 1 to 2 Gy/min at a distance of 80 cm from the teletherapy source. Often the output of a teletherapy machine is stated in Rmm (roentgens per minute at 1 m) as a rough guide to the machine's source strength.

The housing for the source in a teletherapy machine is called the source head. It consists of a steel shell with lead for shielding and a mechanism for bringing the source in front of the collimator opening to produce the clinical gamma-ray beam. Two different methods are in use for moving the teletherapy source from the BEAM OFF into the BEAM ON position and back: (1) a source in a sliding drawer and (2) a source in a rotating cylinder. Both methods incorporate a safety feature that terminates the beam (i.e., moves the source into the OFF position) automatically in case of power failure or emergency. When the source is in the BEAM OFF position, a light appears in the BEAM ON position above the collimator opening, allowing an optical visualization of the radiation field, as defined by the machine collimators and any special shielding blocks.

Some radiation will escape the unit even when the source is in the BEAM OFF position. This head leakage typically amounts to less than 0.01 mSv/hr (1 mR/hr) at 1 m from the source. International regulations require that the average leakage of a teletherapy machine head be less than 0.02 mSv/hr (2 mR/hr) at 1 m from the source.

Linacs as Alternatives to Cobalt-60 Teletherapy Machines

Linacs were developed after 1945 concurrently by two groups: W.W. Hansen's group at Stanford University in the United States (Ginzton et al., 1948) and D. D. Fry's group at Telecommunications Research Establishment in the United Kingdom (Fry et al., 1947). Both groups were interested in linacs for research purposes, used 3,000 MHz as the design frequency, and profited heavily from the microwave radar technology developed during World War II.

The potential for use of linacs in radiation therapy became apparent in the 1950s, and the first clinical linac was installed at the Hammersmith Hospital in London (Miller, 1953). During subsequent years, the linac eclipsed the cobalt-60 machine and became the most widely used radiation source in modern radiotherapy, with several thousand machines in clinical practice around the world today. In contrast to a cobalt-60 unit that provides only two gamma rays with energy of 1.17 MeV and 1.33 MeV (average, 1.25 MeV), a medical linac accelerates electrons to megavoltage kinetic energies in the range from 4 to 25 MeV and can produce two types of radiation beams for use in external beam radiotherapy: electron beams and x-ray beams. Electron-beam radiotherapy linacs accelerate electrons to megavoltage energies, extract them from the accelerating waveguide and transporting them through a beam-forming network to produce a clinical electron beam. X-ray–beam radiotherapy uses an accelerated beam to strike a target in which some of the kinetic energy is transformed into photons. As is explained in Chapter 4, these photons, referred to as bremsstrahlung x-rays, are subsequently formed into a clinical x-ray beam with the help of a flattening filter and special collimators.

In a linac, the electrons are accelerated following straight trajectories in special evacuated structures called accelerating waveguides. Electrons follow a linear path through the same relatively low potential difference a large number of times; hence, linacs fall into the class of cyclic accelerators just like other cyclic machines that provide curved paths for the accelerated particles (e.g., betatrons, microtrons, synchrotrons). The high-power radiofrequency fields, used for electron acceleration in the accelerating waveguide, are produced by devices called magnetrons and klystrons.

Various types of linac are available for clinical use. Some provide x-rays only in the low megavoltage energy range (4 or 6 MV), and others provide both x-rays and electrons at various megavoltage energies (see, e.g., Figure 7-2): typically two photon energies (6 and 18 MV) and several electron energies (e.g., 6–22 MeV). In comparison with cobalt-60 teletherapy machines, linacs are much more versatile for radiotherapy, and their only disadvantages are that they are significantly more expensive to purchase, operate, calibrate, and maintain (costs are discussed later in this chapter). Linacs are usually mounted isocentrically and the operational systems are distributed over five major and distinct sections of the machine:

1. gantry,
2. gantry stand or support,
3. modulator cabinet,
4. patient support assembly (treatment couch),
5. control console.

Cobalt-60 Teletherapy Machine Versus Clinical Linac

In comparison to cobalt-60 machines, linacs have become very complex in design because of:

1. multimodality capabilities (see IGRT, below) that have evolved and are available on most modern machines;
2. increased use of computers in the control systems of these machines;
3. added features, such as high dose rate modes, multileaf collimation, electron arc therapy, and dynamic motion (i.e., while the beam is ON) of the collimators (dynamic wedge), the leaves of a multileaf collimator (IMRT), and the gantry and couch (dynamic stereotactic radiosurgery).

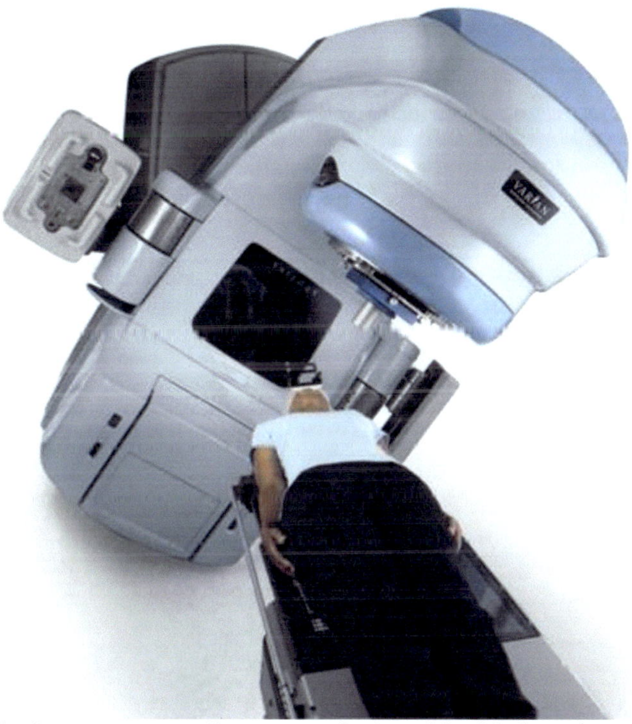

FIGURE 7-2 Modern dual x-ray energy linac manufactured by Varian; the gantry, gantry stand, and the patient support assembly are shown. SOURCE: Image courtesy of Varian Oncology Systems, Palo Alto, CA.

Despite the clear technological and practical advantages of linacs over cobalt-60 machines, the latter still occupy an important place in the radiotherapy toolbox, mainly because of considerably lower capital, installation, and maintenance costs of cobalt-60 machines compared to linacs. In the developing world, the cobalt-60 machines, because of their relatively lower costs, simplicity of design, and ease of operation, are likely to play an important role in cancer therapy for the foreseeable future. Many modern features of linacs, such as multileaf collimators, dynamic wedges, and dynamic operation, can also be installed on modern cobalt-60 machines to allow, at a lower cost, a sophistication in treatment similar to that of linacs. Manufacturers of cobalt-60 machines have been slow in reacting to new technological developments in radiotherapy, conceding preeminence to linac manufacturers even in jurisdictions that would find it much easier and more practical to run cobalt-60 machines as compared to linacs.

Economic Considerations: Megavoltage Linac Versus Cobalt-60 Teletherapy Machine for Standard Radiotherapy

Cobalt-60 machine (isocentric, no beam stopper)

Capital cost: **$750,000**

Operating expense[2]: **$50,000**/year including.

Equipment-related costs only, consisting of (1) maintenance and servicing of equipment (estimated at 4% of capital cost per year = **$30,000/year**) PLUS (2) cost of source replacement every 5 years (new source cost: $ 100,000 or **$20,000/year**)

Output calibration: Every six months.

Linac (low-energy, isocentric, single photon: 6 MV)

Capital cost: **$2,250,000**

Operating expense: **$150,000**/year (equipment only)

Equipment-related cost only, consisting of maintenance and servicing of equipment through either a service contract (estimated at 6.7% of capital cost) or in-house maintenance engineering crew (also estimated at 6.7% of capital cost).

Output calibration: twice per week, that is, 104 times per year.

Linac (high-energy, isocentric: 6 and 18 MV, and 5 electron energies)

Capital cost: **$4,000,000**

Operating expense: **$300,000**/year (equipment only)

Equipment-related cost only, consisting of maintenance and servicing of equipment through either a service contract (estimated at 7.5% of capital cost) or in-house maintenance engineering crew (also estimated at 7.5% of capital cost).

Output calibration: Twice per week.

One additional expense difference is the disposition of the spent cobalt-60 source upon decommissioning of the device. As noted in Chapter 2, some source manufacturers and

[2] Note that the operating expenses provide an estimate for only the annual cost of equipment maintenance and servicing and do not include the personnel cost, such as cost of physics calibration, nor do they include the cost of radiotherapists (radiotherapy technologists) who operate the equipment during the dose delivery to patients.

distributors will take back spent cobalt-60 sources for a fee, and use the pellets to balance the activity of newly irradiated pellets in the manufacture of new sources. Others simply dispose of them. The cost cited above for source replacement includes the cost of removal of the spent source when a new source is purchased. The cost to the owner of the radiotherapy device for disposal of the last source at decommissioning is in the range of $20,000 to $80,000, depending on the condition and origin of the source.

STEREOTACTIC RADIOSURGERY

From an obscure irradiation technique practiced in the 1960s and 1970s in only a few specialized centers around the world, stereotactic irradiation has during the past 20 years developed into a mainstream technique practiced in most major radiotherapy centers around the world. Stereotactic irradiation is the term used to describe focal irradiation techniques that use multiple, non-coplanar radiation beams and deliver a prescribed dose of ionizing radiation to preselected and stereotactically localized lesions primarily in the brain, although progress has been made recently in extending the technique to other parts of the body.

As in standard radiotherapy, the use of multiple beams aimed at the targeted tumor from different directions concentrates the radiation dose in the target and leaves the surrounding tissues with a relatively lower dose. In standard radiotherapy these multiple beams are usually coplanar; in radiosurgery they are non-coplanar.

The main characteristics of stereotactic irradiation are:

1. Total prescribed doses are of the order of 10–50 Gy and the planning targets are small with typical volumes ranging from 1 to 35 cm^3.
2. The requirements for positional and numerical accuracy in dose delivery are ±1 mm and ±5 percent, respectively.
3. Treatment usually is delivered in a single dose or a reduced number of fractions.

The dose in stereotactic irradiation may be delivered through a stereotactic implantation of radioactive sources (stereotactic brachytherapy) or, more commonly, with one or several external radiation sources (stereotactic external beam irradiation).

With regard to dose fractionation, stereotactic external beam irradiation is divided into two categories:

1. Stereotactic radiosurgery: Total dose is delivered in a single session.
2. Stereotactic radiotherapy: Similarly to standard radiotherapy, the total dose is delivered in multiple fractions, often fewer in number than in standard external beam radiotherapy.

From a technical point of view, there is essentially no difference between stereotactic radiosurgery and stereotactic radiotherapy, and often the term radiosurgery is used to describe both techniques. Essentially any radiation beam that has been found useful for external beam radiotherapy has also found use in radiosurgery (cobalt-60 gamma rays, megavoltage x-rays, proton and heavy charged particle beams, and even neutron beams).

Equipment Used for Stereotactic Radiosurgery and Stereotactic Radiotherapy

In addition to a suitable radiation source, in contrast to standard radiotherapy, stereotactic irradiation requires sophisticated specialized equipment and techniques as well as more stringent quality assurance measures. The general list of the specialized equipment is as follows: stereotactic frame, imaging equipment, target localization software, and three-dimensional treatment planning system.

- **Stereotactic frame** defines a fixed coordinate system for an accurate localization and irradiation of the planning target volume. In addition, the stereotactic frame is also used for patient setup on the treatment machine and for patient immobilization during the actual treatment procedure. Figure 7-3 shows two commercial stereotactic frames.
- **Imaging equipment** (computed tomography (CT), magnetic resonance (MR), digital subtraction angiography) is used for visualization, definition, and localization of the structures, lesions and planning target volumes.
- **Target localization software** is used in conjunction with the stereotactic frame system and imaging equipment to determine the coordinates of the target in the stereotactic frame reference system.
- **Treatment planning system** is used for calculation of three-dimensional dose distributions for the radiosurgical treatment. The three-dimensional dose distribution is superperimposed on the patient's anatomical information.

The combined use of stereotaxy and irradiation in treatment of certain brain diseases was introduced in the early 1950s by Swedish neurosurgeon Lars Leksell who also coined the term radiosurgery to describe the technique (Leksell, 1951). Leksell initially used 200 kVp x-rays to deliver, in a single session, a high radiation dose (of the order of 100 Gy) to an intracranial target. He approached the target from several directions to focus the dose on the target within the brain and spare the surrounding vital structures.

Radiosurgery based on kilovoltage x-rays was discontinued in the late 1950s but the idea of focal brain irradiation was carried over to other, more suitable radiation beams, first to protons from cyclotrons (Larsson et al., 1958; Lawrence et al., 1962; Kjellberg et al., 1968) then to focused cobalt-60 gamma rays (Leksell, 1968), and more recently to megavoltage x-rays from linacs (Lutz et al., 1988; Podgorsak et al., 1987; Colombo et al., 1985; Hartmann et al., 1985; Betti and Derechinsky, 1984).

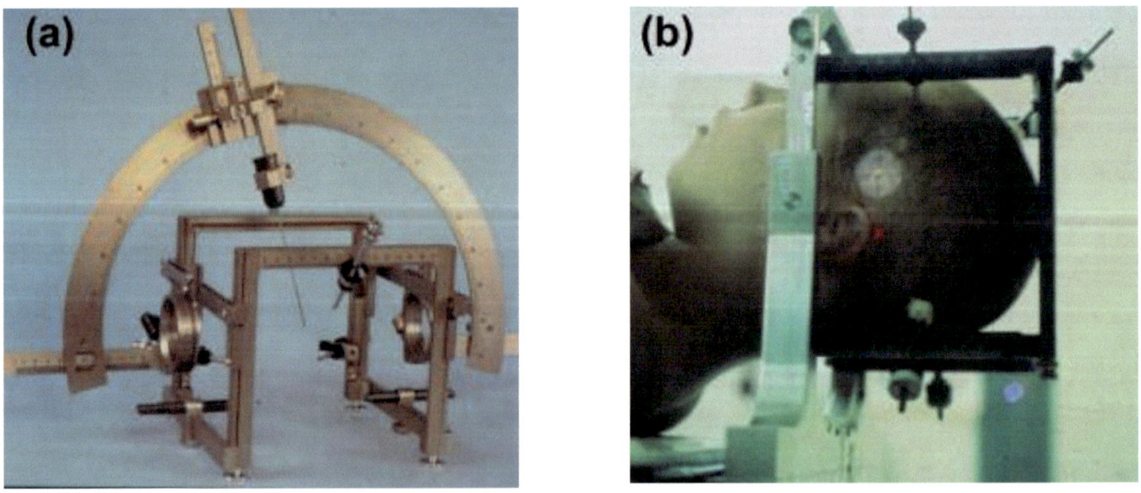

FIGURE 7-3 Two commercial stereotactic frames: (a) Leksell stereotactic frame with attachments for biopsy and (b) OBT frame attached to patient in preparation for linac-based radiosurgery. SOURCE: Images provided by the committee.

In 1974, Larsson proposed linacs as viable radiation sources for radiosurgery (Larsson et al., 1974). In 1984, Betti and Derechinsky from Buenos Aires reported on the development and clinical application of the linac-based multiple non-coplanar converging arcs technique. Soon thereafter in 1985, Colombo and colleagues introduced the technique clinically in Vicenza (Italy) while Hartmann and colleagues introduced it in Heidelberg, Germany.

In 1986, Harvard University in Boston and McGill University in Montreal were the first two institutions to use linac-based radiosurgery in North America. Harvard adopted the multiple non-coplanar converging arcs technique (Lutz et al., 1988), while McGill developed its own radiosurgical technique, referred to as dynamic stereotactic radiosurgery (Podgorsak et al., 1987).

Gamma Knife®

The Gamma Knife® (Elekta, Stockholm, Sweden) is a radiosurgical device that has been associated with, and dedicated to, radiosurgery for the past 40 years. Despite great technological advances during this time, the fundamental design and principles of the Gamma Knife® have not changed much since the Swedish neurosurgeon Lars Leksell introduced the prototype unit in 1968 (Leksell, 1968). The unit incorporates 201 cobalt-60 sources housed in the central body of the unit. These sources produce 201 collimated beams directed to a single focal point (machine isocenter) at a source-focus distance of about 40 cm. The final definition of the circular-beam field size is provided by one of four helmets delivering circular fields with nominal diameters between 4 and 18 mm at the machine focal point (isocenter).

The main components of the Gamma Knife® (see Figure 7-4) are:

- radiation unit with upper hemispherical shield and central body;
- operating table and sliding cradle;
- set of four collimator helmets providing circular beams with diameters of 4, 8, 14, and 18 mm at the isocenter;
- control unit.

Each of the Gamma Knife® cobalt-60 sources is in the form of a steel capsule with a diameter of 1 mm and a height of 20 mm, containing 20 cobalt-60 pellets. The capsule is inserted into another steel capsule, which is enclosed by a bushing and loaded into the central body of the machine. Each source bushing assembly is aligned with its precollimator (6.5 cm of tungsten alloy), stationary collimator (9.25 cm of lead), and the final collimator (6 cm of tungsten alloy) on one of the four helmets.

A newly loaded Gamma Knife® has a total activity of the order of 222 TBq (6,000 Ci) and all individual source activities are within 5 percent of an average source activity, which is of the order of 1.11 TBq (30 Ci). The dose rate at the center of a spherical water-equivalent phantom with a radius of 8 cm placed with the sphere center into the isocenter of the Gamma Knife® is on the order of 3 Gy/min. This rate will decrease to 50 percent of its original value during one half-life of the cobalt-60 radionuclide (5.26 years).

A Chinese company, GammaStar® Medical Group, Ltd., is introducing its own cobalt-60 radiosurgery device, called Gyro Knife. This device is the functional equivalent of the isocentric linac dedicated to stereotactic radiosurgery, described below. The Gyro Knife therapy head pivots the 220- to 260-TBq (6,000- to 7000-Ci) source on an axis, using a multileaf collimator to direct the radiation beam at the tumor, while the gantry rotates the therapy head around the patient's body. Unlike Gamma Knife®, this device can be used for tumors in any part of the body.

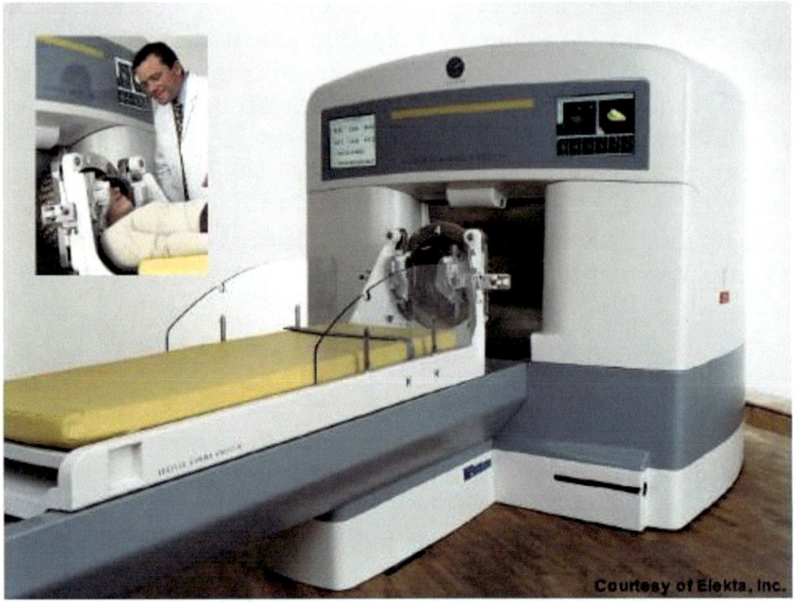

FIGURE 7-4 A Gamma Knife® installation showing the main body of the unit containing 201 cobalt sources (at 30 Ci = 1.11 TBq each source), the treatment couch, and a collimator helmet attached to the treatment couch. The inset shows an up-close image of the automatic positioning system used to position the patient for treatment. SOURCE: Image courtesy of Elekta.

Nonradionuclide Replacements for Cobalt-60 Radiosurgery

There are currently three options for nonradionuclide alternatives to cobalt-60 radiosurgery: Stereotactic radiosurgery based on a standard isocentric linac, stereotactic radiosurgery with dedicated isocentric linac, and a miniature linac on a robotic arm (CyberKnife®).[3] Each of these is discussed below.

Stereotactic radiosurgery based on a standard isocentric linac

In contrast to Gamma Knife® which is dedicated solely to stereotactic radiosurgery, linac-based radiosurgery uses a standard isocentric linac with tight mechanical and electrical tolerances, modified for radiosurgery. This means that radiosurgery can be performed on such a linac on top of the daily routine radiotherapy patient load. The required modifications to a standard linac for use in radiosurgery consist of:

- Supplementary collimation, either in the form of a set of collimators to define the small diameter circular radiosurgical beams or a micro-multileaf collimator (micro-MLC), to define the small area irregular fields.
- Remotely controlled motorized couch or treatment chair rotation.
- Couch brackets or a floor stand for immobilizing the stereotactic frame, i.e., the patient, during treatment.
- Inter-locked readouts for angular and height position of the couch.
- Special brakes to immobilize the vertical, longitudinal, and lateral couch motions during treatment.

[3] One other technique, proton beam radiation therapy (also called proton therapy), can also be used to treat the same tumors that are treated with a GammaKnife® but at two or three times the expense. Proton therapy is in some ways superior to standard radiation therapy techniques, but is not discussed here because of the clear cost difference from linac and gamma radiosurgery.

Isocentric linac-based radiosurgical techniques currently fall into two categories: (i) multiple non-coplanar converging arcs and (ii) dynamic stereotactic radiosurgery. Each linac-based technique is characterized by a particular set of individual rotational motions of the linac gantry and the patient support assembly (couch or chair) from given start to stop angles. Of the two approaches, the multiple converging arcs technique is the more common.

In the multiple non-coplanar converging arcs technique the patient is stationary on either the treatment couch or chair, while the gantry moves through a given arc. In the dynamic stereotactic radiosurgery technique, both the gantry and the patient rotate simultaneously during the dose delivery (the gantry moves 300 degrees, from 30 to 330 degrees, and the couch moves 150 degrees, from −75 to +75 degrees; see Figure 7-5).

Stereotactic Radiosurgery with Dedicated Isocentric Linac

Lately, linacs dedicated solely to stereotactic radiosurgery have become commercially available. Some still use stereotactic frames and operate essentially in the same manner as the standard linacs modified for radiosurgery; others (e.g., Novalis®, BrainLab) actually dispense with the stereotactic frame and achieve high precision without a frame using a mask for immobilization and on-line x-ray imaging of internal structures. Orthogonal stereoscopic x-ray sources are placed below the floor level in the treatment room, and ceiling-mounted amorphous silicon flat-panel detectors provide diagnostic-quality imaging of the patient anatomy in the treatment position. These sophisticated frameless techniques use a micromultileaf collimator for shaping of irregular fields and allow intracranial as well as extracranial stereotactic irradiation with equipment and operating costs similar to those for a Gamma Knife®. Figure 7-6 shows a linac (Novalis®, BrainLAB) installation dedicated to stereotactic radiosurgery.

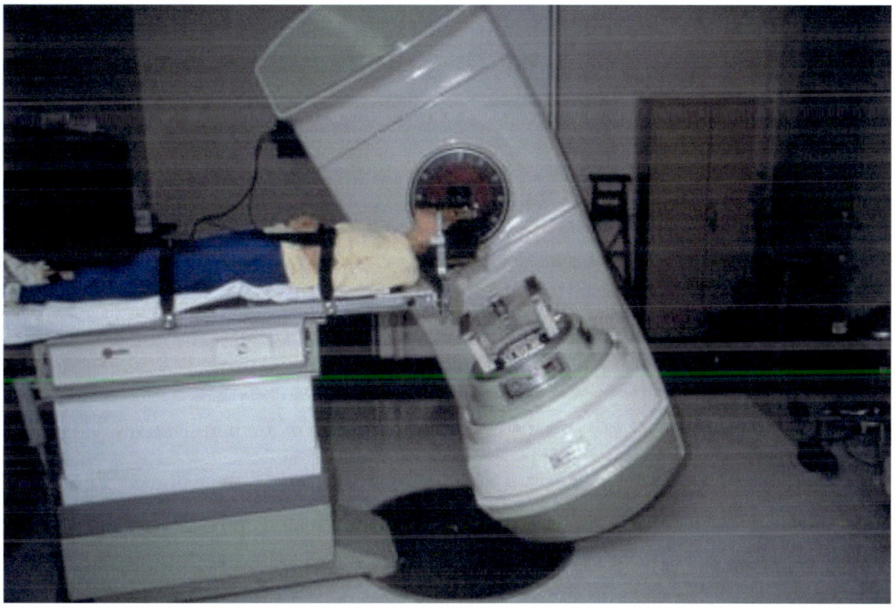

FIGURE 7-5 Patient receiving linac-based dynamic stereotactic radiosurgery on a Clinac 18, a linac manufactured in the 1970s by Varian Medical Systems. More recent medical linacs have been developed specifically for stereotactic radiosurgery incorporate image-guidance and beam-shaping technologies.

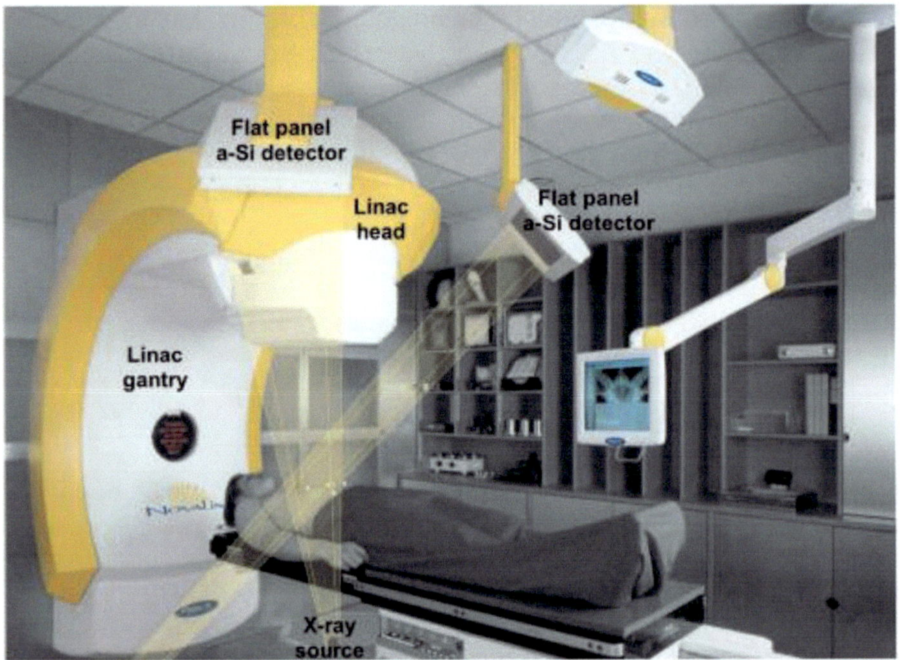

FIGURE 7-6 Linac installation dedicated to stereotactic radiosurgery (Novalis®, BrainLAB). The isocentric linac operates at 6 MV; imaging is carried out with two x-ray sources installed below the treatment room floor and two ceiling-mounted amorphous silicon flat-panel detectors. SOURCE: Image courtesy of BrainLAB; Heimstetten, Germany.

Miniature Linac on a Robotic Arm (CyberKnife®)

The CyberKnife® radiosurgery system (see Figure 7-7) provides a different approach to image-guided dose delivery utilizing an on-line orthogonal pair of digital x-ray imagers, a patient CT data set fused with MR and/or positron emission tomography images and a miniature linac mounted on an industrial robotic arm. This broadens the range of traditional stereotactic radiosurgery: Because it monitors and tracks the patient position continuously and uses on-line images for finding the position of the target in the treatment room coordinate system, the CyberKnife® allows frameless radiosurgery; that is, it can operate without a rigid and invasive stereotactic frame. The device directs the radiation beam into the target with a reported dose-delivery accuracy on the order of 1 mm. Because the linac is mounted on a robotic arm, it also can perform frameless radiosurgical dose delivery to extracranial targets, such as the spine, lung, and prostate, by using the body skeleton or surgically implanted fiducial markers as frame of reference for targeting purpose.

Gamma Knife® Versus Linac-Based Radiosurgery

The introduction of linac-based radiosurgery in radiation oncology departments during the late 1980s has very rapidly transformed radiosurgery from an obscure technique practiced in only a few specialized neurosurgery departments around the world into a mainstream radiotherapeutic technique. This stimulated great advances in technical and clinical utility of radiosurgery. However, the introduction of radiosurgery into radiation oncology departments has also caused some problems and differences of opinion between neurosurgeons, who were the inventors and until then the principal users of radiosurgery, and radiation oncologists, who are the professionals trained and licensed in treatment of disease with ionizing radiation and are quite comfortable with the clinical use of isocentric linacs.

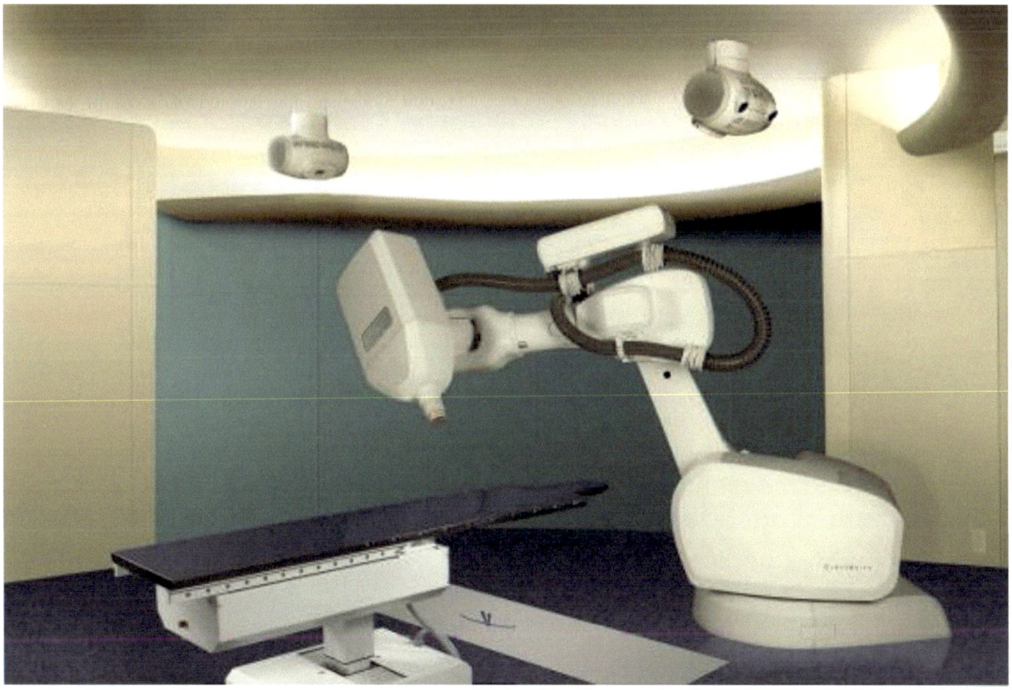

FIGURE 7-7 CyberKnife® installation consisting of miniature linac mounted on a robotic arm, treatment couch, and imaging device: ceiling-mounted x-ray tubes and floor-mounted image intensifiers. SOURCE: Image courtesy of Accuray, Sunnyvale, CA.

Radiation oncologists, on the one hand, embraced the new linac-based radiosurgical techniques with great enthusiasm, but had some reservations about the use of single high-dose irradiation in radiosurgery in contrast to the multifractionated schemes used in conventional radiotherapy. The neurosurgeons, on the other hand, have had previous favorable experience with Gamma Knife® radiosurgery and have expressed concern about the mechanical stability of isocentric linacs when used in radiosurgery.

An unstable linac isocenter could adversely affect the accuracy of dose delivery and result in substandard treatments in comparison to treatments provided by the 201 stationary beams from the Gamma Knife®. These concerns are valid, and clearly not all isocentric linacs are suitable for conversion to radiosurgery. However, there is no question that a well-designed, well-aligned, and properly maintained isocentric linac will have a stable and small enough isocenter sphere (of the order of 1-mm diameter) making it suitable for use in radiosurgery.

The debate on the relative merits of Gamma Knife® versus linac-based radiosurgery continues, but one thing is clear: the Gamma Knife® incorporates some 200 cobalt-60 sources, each with an activity of 1.11 TBq (30 Ci), whereas linac-based radiosurgery does not use any radioactive material.

The general consensus among radiation oncologists and medical physicists is that linac-based radiosurgical treatments with regard to treatment outcomes are equivalent to those provided by the Gamma Knife®. However, linac-based techniques, in comparison, with Gamma Knife® techniques, are more complicated and slower, but have greater potential for new technical and clinical developments, such as intensity modulation, extracranial application, on-line imaging, fractionated as well as image-guided dose delivery, and automatic patient repositioning.
Informal examination of the usage and publications in this area suggests that the majority of neurosurgeons and some radiation oncologists believe that the Gamma Knife® is superior to any linac-based radiosurgical technique. During the past decade this apparent consensus has resulted in over 100 Gamma Knife® installations in the United States, many of them installed in neurosurgery departments.

Economic Factors: Radiosurgery with Gamma Knife® Versus Radiosurgery with Isocentric Linac

The introduction of linac-based stereotactic radiosurgery techniques in the 1980s has not only stimulated rapid growth in clinical radiosurgery, it has also started a heated debate on the clinical and economic merits of the linac-based versus Gamma Knife® approach. A comparison among the three commonly used approaches to stereotactic radiosurgery (Gamma Knife®, dedicated linac, and modified standard linac) is as follows:

Gamma Knife®
- Capital cost (machine): $4,000,000
- Capital cost (bunker): $2,000,000
- Annual operating cost (equipment and infrastructure only):
  - Amortization (10 percent of capital cost): $600,000
  - Service contract (8 percent of equipment capital cost): $320,000
  - Source exchange (20 percent of cost at 5 years): $200,000
- **Total annual operating cost (infrastructure and equipment): $1,120,000**

Radiosurgery with a dedicated linac
- Capital cost (machine): $4,000,000
- Capital cost (bunker): $2,000,000
- Annual operating cost (equipment and infrastructure only):
  - Amortization (10 percent of capital cost): $600,000
  - Service contract (8 percent of equipment capital cost): $320,000
- **Total annual operating cost (infrastructure and equipment): $920,000**

Radiosurgery based on modified standard isocentric linac
- Capital cost (machine): machine already available for standard radiotherapy
- Capital cost (bunker): already available for standard radiotherapy use
- Cost of linac modification for radiosurgery: $50,000
- Cost of micro-MLC and radiosurgical treatment planning system: $600,000
- Annual operating cost (radiosurgical equipment only):
  - Amortization (10 percent of modification cost): $65,000
  - Service contract (8 percent of $600,000): $48,000
- **Total annual operating cost (equipment): $113,000**

Just as in the comparison of radiotherapy costs, an additional expense difference is the decommissioning cost of the Gamma Knife®, including disposition of the last set of spent cobalt-60 sources. This cost is estimated to be $50,000 to $70,000.

Undoubtedly, the clinical utility of the Gamma Knife® is well proven; however, radiosurgery based on the Gamma Knife® is also significantly more expensive than that based on a modified linac, assuming, of course, that the linac is used for standard radiotherapy and carries radiosurgery as additional load to the standard treatment load.

Departments with large neurosurgical patient loads require dedicated stereotactic radiosurgery equipment and then a choice between a Gamma Knife® and dedicated linac must be made. In this situation, the costs for infrastructure, equipment, servicing, and operation of a Gamma Knife® and dedicated linac are very similar, except that the Gamma Knife® costs must also include expensive cobalt-60 source replacement after every 5 years of operation and eventual decommissioning costs for disposal of the last set of sources.

VETERINARY RADIOTHERAPY

Radiotherapy is also used in veterinary medicine, essentially with same equipment as that used in human radiation oncology. The U.S. Veterinary Cancer Society lists about 70 therapy machines in clinical service. There are no cesium-137 teletherapy machines on the list of veterinary service machines, although it is possible that a veterinary school still has one, but the schools are not included in this tally.

FINDINGS

In the United States and other developed countries, cobalt-60 teletherapy machines have during the past two decades largely been supplanted by linacs which are considered a better and more versatile tool for provision of standard and advanced radiotherapy. In developing countries, cobalt-60 teletherapy machines will likely continue to play a major role in radiotherapy because of their significantly lower capital and operating costs in comparison with linacs.

Radiosurgery is a special irradiation technique practiced mainly with the Gamma Knife® which uses around 200 cobalt-60 sources in combination. The use of Gamma Knife® for radiosurgery is not essential, because the technique can also be practiced with isocentric linacs as well as with a miniature linac mounted on a robotic arm. However, many neurosurgeons and some radiation oncologists believe that Gamma Knife® radiosurgery is superior to linac-based radiosurgery. The cost of radiosurgery with a dedicated linac is similar to that with a Gamma Knife®, whereas the cost of radiosurgery with a standard linac modified for radiosurgery is only a small fraction of that incurred with a Gamma Knife®. In addition, linac-based radiosurgery has greater potential than the Gamma Knife® for development of stereotactic techniques in the treatment of small localized lesions not only in the brain but also in other organs of the human body and not only on a single treatment basis but also on a fractionated basis, both under daily image guidance.

Development of new treatment technologies and shifts in practices in radiotherapy are driven by innovation and decisions about delivery of health care, both at the macrolevel (e.g., insurance reimbursements) and at the personal level (i.e., doctor-patient treatment planning). Medical technology researchers seek out opportunities to improve nonradionuclide-based radiotherapy and radiosurgery. Because cesium-137–based teletherapy is no longer practiced in the United States and there are only a few cobalt-60 teletherapy machines left in clinical operation in the United States, the committee concludes that teletherapy services, while important, are lesser and declining concerns with respect to radiation source security. It is also possible that the recent rapid growth in Gamma Knife® installations in the United States and the developed world will soon subside given that a viable alternative is offered with linac-based radiosurgery. An important issue is the end-of-life source disposal, which currently is somewhat costly and cumbersome. This encourages storage of spent machines including Co-60 sources in places that may be poorly secured.

CHAPTER 8

INDUSTRIAL RADIOGRAPHY

SUMMARY

Gamma-ray radiography is one of a number of technologies used in industry for safety assessment and quality control purposes. In particular, it is widely used in the chemical, petrochemical, and building industries for radiographic inspection of pipes, boilers, and structures where the economic and safety consequences of failure can be severe. With ongoing developments in ultrasonic inspection technologies as well as in x-ray radiography, satisfactory alternatives for many of the gamma-ray radiography applications exist. There are, however, a number of specific applications, such as pipeline inspection in remote locations, underwater, and in chemical plants with wrapped or closely spaced piping, where the advantages of mobility, ease of use, and low power requirements of radionuclide-based radiography may make replacement difficult and uneconomic, at least with current alternative technologies.

Although gamma-ray radiography is generally performed with Category 2 iridium-192 and cobalt-60 sources, their portability and use in remote locations poses higher exposure risks than other, equivalent-activity radionuclide sources. Improved detector technology could enable the use of lower activity sources with consequently decreased risk.

The committee judges that except in some specialized applications, alternative inspection technologies are already increasingly replacing gamma-ray radiography in industry. In some instances, such as ultrasonic inspection, the replacement rate is currently limited by the availability of trained personnel.

NONDESTRUCTIVE INSPECTION

Radiography, using radionuclide sources and x-ray sources, is one of a number of techniques developed for the nondestructive inspection (NDI) of structures, such as pipes, and components manufactured by industry today. Radiography and other NDI techniques have become indispensable in many industrial sectors for safety assurance purposes because they have the capability to detect the presence of manufacturing defects, such as incomplete welds and porosity in the walls of tubes, as well as cracks and other flaws that develop during service. Because defects can compromise safety and lead to failure, NDI is governed by standard codes of practice in many industrial settings and may be mandated by law. For instance, pipelines, such as the Alaska pipeline, are constructed from welded sections, and the welds joining them are inspected during construction and periodically during their operating lives to check for cracks and corrosion. Similarly, pipes in chemical plants and refineries are routinely inspected during the construction of the plant and periodically during service. The construction industry also relies on NDI when building with steel-reinforced concrete: The integrity of rebar reinforcements during building construction is usually inspected to ensure that they are continuous and intact. In fact, NDI by radiography is practiced in an enormous range of applications and industries and has grown to be a major industry in its own right.

Radiation sources are used widely for examining gas and oil pipelines, pipes and pressure vessels in chemical plants, vehicles, and aircraft, and although the radionuclide sources provide some undeniable benefits, they also pose some risks. These are applications in which the convenience, small size, and mobility of radionuclide radiography sources make them particularly attractive to the service industry. The small size and mobility of the sources also make them more vulnerable to seizure than many other types of radioactive sources considered by the committee.

Industrial radiography, whether by x-ray radiography or gamma-ray radiography, is used to detect defects, for instance, in much the same way as dentists use x-ray radiography to check for decay in teeth. Transmission of radiation through a component produces an image on film that is readily interpretable, revealing spatial and density variations. These images are traditionally recorded on sheets of film, although increasingly film is being replaced by solid-state detectors. The process is illustrated in Figure 8-1. When there is a discontinuity in the material, for example, a void or crack, the radiation absorption through that section is reduced and more radiation passes through the material, giving rise to contrast on a film or a variation in signal on a detector. Indeed, radiography is often referred to as a volumetric inspection because it gives a projected picture of the internal structure of a component.

The thickness of a component that can be inspected and through which an image can be recorded depends on the shielding properties of the component material and the energy and intensity of the radiation. The shielding properties depend on the atomic mass of the component material and the energy of the radiation. Lead shields radiation very strongly (it has a high atomic number), allowing inspection of only thin components. Concrete is a much less effective shield for gamma and x-rays because it is composed of lower-atomic-number elements and so can be examined in thick sections.

Similarly, plastics are even less effective as shielding materials and so still thicker sections can be inspected. Higher energy gamma and x-rays have greater penetrating capabilities. Many components and structures are made of materials such as concrete, steel, and other metals, and quite often composites of these materials. If one uses too much energy for a given application, the film or detector can become saturated or contrast can be lost. Industrial radiographers would like to use sufficiently high energy to penetrate the part and form an image in a reasonable amount of time but not so much to lose contrast (because low energies and high absorption increase contrast) or saturate the detector. The most commonly used sources are iridium-192 and cobalt-60. Iridium-192 emits gamma rays with a range of energies up to 820 keV (averaging 380 keV) and cobalt-60 emits two gamma rays, one at 1.173 MeV and one at 1.333 MeV. The peak energy is important because it corresponds to the most penetrating gamma rays. The typical activity of these sources makes them Category 2. For certain specialized applications, selenium-75 and ytterbium-169 are also used. These radionuclides are listed along with their half-lives and the average energy of their gamma rays in Table 8-1.

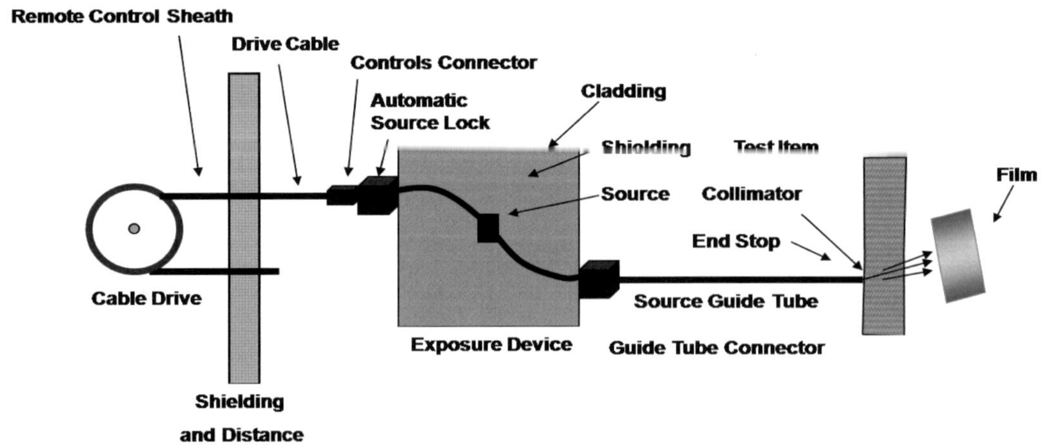

FIGURE 8-1 Schematic of the internals of a portable, hand-held radiography camera. Such a system typically uses iridium-192 as the gamma source. SOURCE: Image provided by the committee.

TABLE 8-1 Properties of Gamma-Emitting Radioactive Isotopes Most Commonly Used in Industrial Radiography

Radionuclide	Average Gamma Energy (keV)	Half-Life
Cobalt-60	1,250	5.3 yr
Iridium-192	380	75 d
Selenium-75	217	120 d
Ytterbium-169	145	32 d

SOURCE: Table provided by the committee.

One parameter widely used in the industry to measure the penetration of gamma- and x-ray radiation having a given energy is the half-value thickness, which is the thickness at which the radiation intensity has decreased to one-half through absorption and scattering processes. The half value is dependent on the energy of the incident irradiation and the density of the material. Figure 8-2 shows the half-value thickness for several common shielding materials.

Radionuclide Radiography Sources

The majority of mobile gamma ray radiography devices use iridium-192 sources, although there are a significant number of radiography devices using cobalt-60 sources. These latter sources require more shielding. Consequently, they are heavier and are not usually hand-held devices but nevertheless are mobile when mounted on a trolley.

A photograph of a hand-held device is shown in Figure 8-3 and the internal structure is shown schematically in Figure 8-1. When the device is not in use, including during transport, the radiation source is contained within the camera in a depleted uranium shield so that external radiation is reduced to safe levels. In operation, to record a radiographic image, the source is moved from the shielded region through a tube to the collimator/end stop, allowing gamma rays to pass through the test item, such as a pipe, exposing the film to form a radiograph. When the radiograph is complete, the source is retracted into the shield and the device moved to make another inspection either in the same location or driven to another site.

Nature of the Industrial Gamma-ray Radiography Industry

The industrial gamma-ray radiography industry is highly diverse and consists of a large number of individual companies. These range from relatively large companies offering a wide variety of NDI tools and services to small operators specializing, for instance, in gamma-ray radiography for in-field pipeline inspection. Furthermore, the relatively low cost of gamma sources, their portability, straightforward image interpretation, and simple radiological safety measures makes entry into the market relatively inexpensive for small companies. Also, the use of portable gamma sources allows radiography to be carried out in remote locations where electric power may not be readily available.

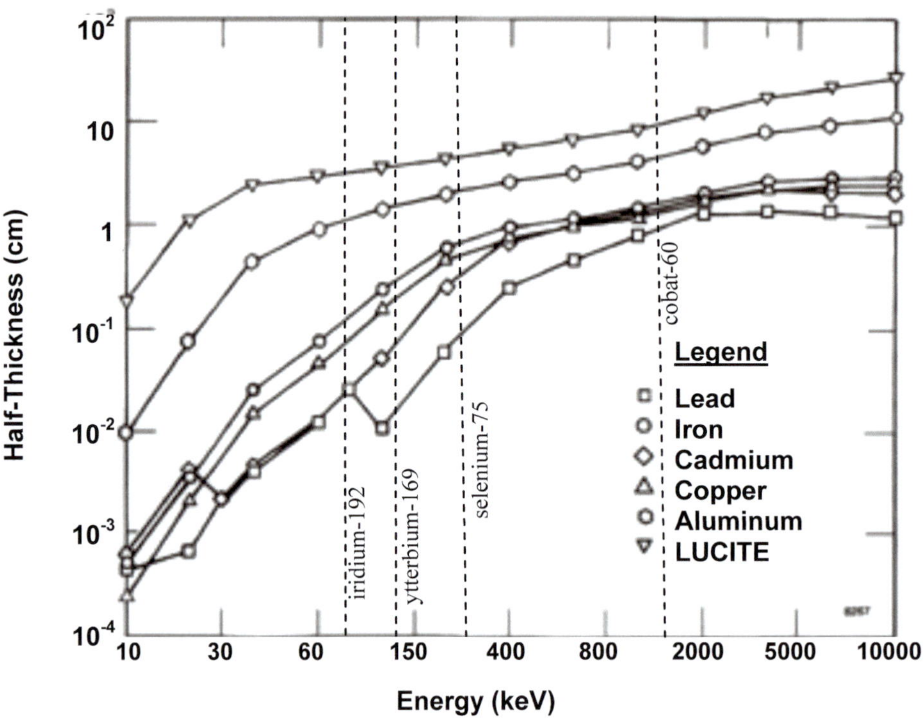

FIGURE 8-2 The half-value thickness (thickness that shields one-half of the incident photons) is shown for several common shielding materials. Each dashed line indicates the approximate location of the average energy for the gamma rays emitted by the radionuclide with which it is labeled. Note that the actual half-value is slightly higher for radionuclides that emit some higher-energy gamma rays, such as iridium-192. SOURCE: Adapted from ORTEC (2007).

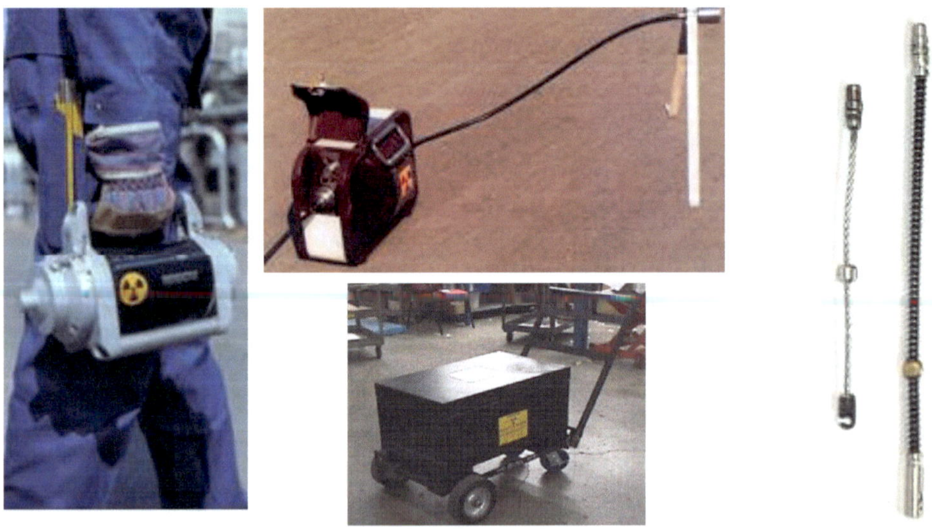

FIGURE 8-3 Radiography cameras. Note that the hand-held device is an iridium-192 radiography camera, as is the sealed source mounted on the end of the cable shown to the right. The device on wheels is a cobalt-60 radiography camera. SOURCE: Images provided by the committee.

Safety Issues

During use, when the source is outside its storage shielding, it can create radiation fields in which permissible occupational dose standards can be exceeded in a short period of time. Data indicate that occupational exposures received by gamma radiography workers are among the highest of all radiation workers in the United States. The average total effective dose equivalent (TEDE) for personnel working in gamma radiography in 2005 was 5.2 mSv (520 mrem), compared to 1 mSv (100 mrem) for workers in commercial nuclear power reactors. If the devices are not handled properly, the radiation sources can cause radiation overexposures and radiation burns. Accidents occur primarily when the radiographer does not return the source to the fully shielded position and fails to perform a survey to confirm the source is in a fully shielded position. Those performing radionuclide radiography are subject to much greater radiation exposures than those in the medical irradiator sectors, where exposures are not routine parts of the job.

Occupational doses to workers using radiation generators are much lower. Radiation generators are typically placed in shielded rooms or bunkers and can be turned off in an emergency situation. Further, portable radiation generators can be turned off when they are not in use.

The safety concerns associated with industrial radiography also have economic ramifications. For instance, the immediate vicinity of locations where gamma radiography is being performed has to be cleared of other personnel. At sites where there are other workers, it is not unusual for other workers to have to stop work while the radiography is carried out. This is a cost and one of the factors that can influence the choice of nonradioactive methods of NDI because for some alternatives the area does not need to be cleared and other work can continue uninterrupted.

Regulatory Framework

There are two regulatory frameworks that govern industrial radionuclide-based radiography. One governs the use (through licensing) and the other governs the transportation of the radiography machines containing radionuclide sources in addition to those described earlier. In addition, transportation of portable radionuclide sources that are used in the field is regulated by Department of Transportation regulations contained in 49 CFR Parts 100-185 and U.S. NRC transportation regulations contained in 10 CFR Parts 20 and 71. Technicians using radioactive sources are certified by either state programs or the American Society for Non-Destructive Testing. These certification programs comply with 10 CFR Part 34—Licenses for Industrial Radiography and Radiation Safety Requirements for Industrial Radiographic Operations.

Because of safety concerns, industrial gamma-ray radiography sources are considered to be a "special form" and must meet "special form requirements": (1) the radionuclide source must be contained in a solid piece or a sealed capsule that can only be opened by destroying the capsule, (2) at least one dimension of the capsule must not be less than 5 mm, and (3) the source must satisfy the specific requirements of 49 CFR § 173.469 that the source capsule does not break, melt, or leak after being subjected to a variety of prescribed impact, deformation, leaching, and high-temperature exposure tests.

The other regulatory framework is much more complex and concerns the accepted use and methodologies of NDI techniques for specific applications. These are embodied in codes that, as in many other industries in the United States, and internationally, cover the design, manufacture, inspection, insurance coverage and qualification, and operation of its systems. For some applications, for instance in assuring that welds are defect-free, permanent retention of the radiographic images is sometimes also a legal requirement.

Many of the codes covering industrial radiography, using radionuclides and x-ray sources, pertinent to the committee's charge have been promulgated by the American Society of Mechanical Engineers (ASME) Boiler and Pressure Vessel Code and by the American Petroleum Institute (API). The former establishes the rules of safety governing the design, fabrication, and inspection of boilers and pressure vessels and nuclear power plant components during construction to provide a margin for deterioration during service. Radiographic examination has been permitted by the ASME Boiler and Pressure Vessel Code since 1931. The API codes govern the manufacture of oil and gas steel pipelines, how they are assembled by welding, and their inspection.

The codes of practice not only prescribe the use of specific inspection techniques but can also affect the choices made in selecting alternative inspection techniques. In some cases, radiography using radionuclide sources is specified, whereas in others radiography is described without specifying whether the source is a radionuclide or an x-ray machine. In the opinion of the committee, changes could be made by the governing bodies of the codes after appropriate representation of the merits and viability of alternative inspection modalities to radionuclide radiography. However, there is also a structural inertia in the adoption of alternative methods associated with the way in which codes are used. This can be illustrated by the hypothetical example in Figure 8-4 in which a refinery plant operator contracts to have its plant's pipes inspected.

In this example, companies that offer inspection services respond with a quotation for the service. Unless specified by the plant operator, the service company typically selects a technique for the inspection based on its expertise and the practices that its insurance carrier will accept. In turn, the methods that the inspection company can employ must comply with the technical codes of practice. The inertia in replacing radionuclide radiography sources with a technically viable alternative largely comes from the time it takes for the technical committees of the code-making organizations to consider and change the code if it decides it is appropriate.

Alternative Technologies

There are many NDI techniques in use today, ranging from eddy current testing, acoustic emission, and radiography to magnetic induction and ultrasonics. Table 8-2 lists commonly used methods of nondestructive testing and their effectiveness in detecting particular kinds of defects, relevant to the inspection of pressure vessels and pipes where radiography has traditionally been a principal tool.

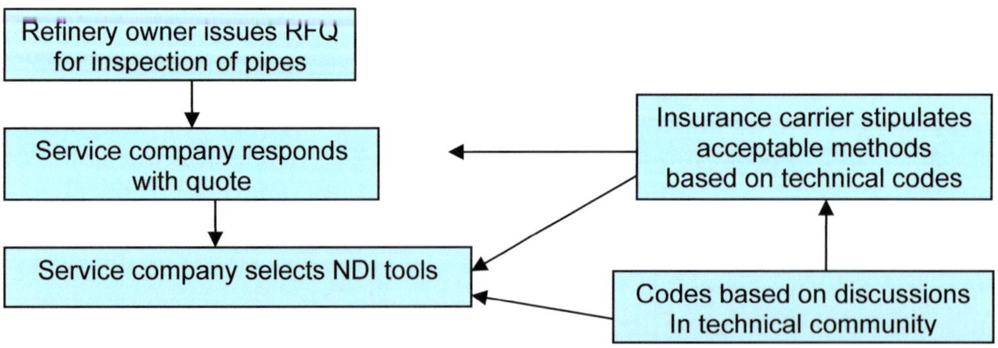

FIGURE 8-4 Process for selecting nondestructive testing method on the left, and factors that influence or dictate the process on the right. NOTE: RFQ = Request for quote. SOURCE: Provided by committee.

TABLE 8-2 Common Imperfections and the Nondestructive Examination Methods That Are Generally Capable of Detecting Them

	Surface [Note (1)]		Sub-surf. [Note (2)]		Volumetric [Note (3)]				
	VT	PT	MT	ET	RT	UTA	UTS	AE	UTT
Service-Induced Imperfections									
Abrasive Wear (Localized)	●	◑	◑		●	◑	◑		◑
Baffle Wear (Heat Exchangers)	●			◑					
Corrosion-Assisted Fatigue Cracks	○	◑	●		○	●		●	
Corrosion -Crevice	●								○
-General / Uniform				○	◑		◑		●
-Pitting	●	●	○		●	○	○	◑	○
-Selective	●	●	○						○
Creep (Primary) [Note (4)]									
Erosion	●				●	○	◑		◑
Fatigue Cracks	○	●	●	◑	◑	●		●	
Fretting (Heat Exchanger Tubing)	◑			◑					◑
Hot Cracking		◑	◑		◑	○		◑	
Hydrogen-Induced Cracking		◑	◑		○	◑		◑	
Intergranular Stress-Corrosion Cracks						○			
Stress-Corrosion Cracks (Transgranular)	○	◑	●	○	◑	◑		◑	
Welding Imperfections									
Burn Through	●				●	◑			○
Cracks	○	●	●	◑	◑	●	○	●	
Excessive/Inadequate Reinforcement	●				●	◑	○		○
Inclusions (Slag/Tungsten)			◑	◑	●	◑	○	○	
Incomplete Fusion	◑		◑	◑	◑	●	◑	◑	
Incomplete Penetration	◑	●	●	◑	●	●	◑	◑	
Misalignment	●				●	◑			
Overlap	◑	●	●	○		○			
Porosity	●	●	○		●	◑	○	○	
Root Concavity	●				●	◑	○	○	○
Undercut	●	◑	◑	○	●	◑	○	○	
Product Form Imperfections									
Bursts (Forgings)	○	●	●	◑	◑	◑	◑	●	
Cold Shuts (Castings)	○	●	●	○	●	◑	◑	○	
Cracks (All Product Forms)	○	●	●	◑	◑	◑	○	●	
Hot Tear (Castings)	○	●	●	◑	◑	◑	○	○	
Inclusions (All Product Forms)			◑	◑	●	◑	○	○	
Lamination (Plate, Pipe)	○	◑	◑			○	●	○	●
Laps (Forgings)	○	●	●	○	◑		○	○	
Porosity (Castings)	●	●	○		●	○	○	○	
Seams (Bar, Pipe)	○	●	●	◑	○	◑	◑	○	

Legend:
AE – Acoustic Emission
ET – Electromagnetic (Eddy Current)
MT – Magnetic Particle
PT – Liquid Penetrant
RT – Radiography
UTA – Ultrasonic Angle Beam
UTS – Ultrasonic Straight Beam
UTT – Ultrasonic Thickness Measurement
VT – Visual

● – All or most standard techniques will detect this imperfection under all or most conditions.
◑ – One or more standard technique(s) will detect this imperfection under certain conditions.
○ – Special techniques, conditions, and/or personnel qualifications are required to detect this imperfection.

GENERAL NOTE: Table A-110 lists imperfections and NDE methods that are capable of detecting them. It must be kept in mind that this table is very general in nature. Many factors influence the detectability of imperfections. This table assumes that only qualified personnel are performing nondestructive examinations and good conditions exist to permit examination (good access, surface conditions, cleanliness, etc.).

NOTES:
(1) Methods capable of detecting imperfections that are open to the surface only.
(2) Methods capable of detecting imperfections that are either open to the surface or slightly subsurface.
(3) Methods capable of detecting imperfections that may be located anywhere within the examined volume.
(4) Various NDE methods are capable of detecting tertiary (3rd stage) creep and some, particularly using special techniques, are capable of detecting secondary (2nd stage) creep. There are various descriptions/definitions for the stages of creep and a particular description/definition will not be applicable to all materials and product forms.

SOURCE: Reprinted from ASME 2004 Boiler and Pressure Vessel Code, Section V, by permission of ASME.

The most direct substitute for radionuclide radiography is the use of x-ray radiography because x-ray sources can produce radiation over a range of different energies, including those generated by the radionuclides, to perform the same inspection. In many industrial applications, such as inspection of high-value components, electronic circuits, and devices (spark plugs, for instance), x-ray radiography is already becoming more widely used as a quality control tool in manufacturing, especially with the development of compact, microfocus x-ray tubes. These industrial applications typically occur in factory settings where components can be inspected in a fixed location and the higher cost of x-ray facilities is outweighed by their increased productivity, improved imaging resolution, or higher energy for greater penetration. Advances in compact, high-energy x-ray sources as well as recent developments in computerized x-ray tomography, which can provide three-dimensional images of a component, promise to extend the use of x-ray radiography to applications outside factory settings. Already a number of companies around the world offer x-ray radiography inspection services in place of gamma-ray radiography for pipe inspection. A number of companies in the United States and elsewhere provide compact linacs and betatrons to replace the radionuclide sources (see Figure 8-5 and Chapter 4).

For inspection of larger components and structures, such as boilers, pipelines, and bridges, other techniques (principally ultrasonic inspection) have been gaining ground on radiography. One of the principal reasons is that radiography only provides a projection image and is relatively insensitive to features such as thin cracks aligned perpendicular to the beam, which give poor contrast. Figure 8-6 illustrates a radiography source in use examining a component that has a thin crack perpendicular to the radiation tracks and a thin crack parallel to the radiation tracks. A cartoon of the exposed film can be found in the lower right-hand corner of the figure, showing that the defect aligned parallel to the gamma-ray paths is detectable and the defect perpendicular to the radiation is not. This is especially important in the inspection of complex-shaped components where defect identification is difficult in the presence of variations in thickness associated with their shape. In contrast, the use of ultrasonics enables images to be formed in different conditions to increase the likelihood of detecting flaws.

FIGURE 8-5 Photograph of an Inspecta betatron particle accelerator hanging in a harness and used for nondestructive examination. The x-ray energy can be varied from 2 MeV up to 6 MeV, which allows the radiographer to examine thick sections of steel or concrete. SOURCE: Copyright Inspecta.

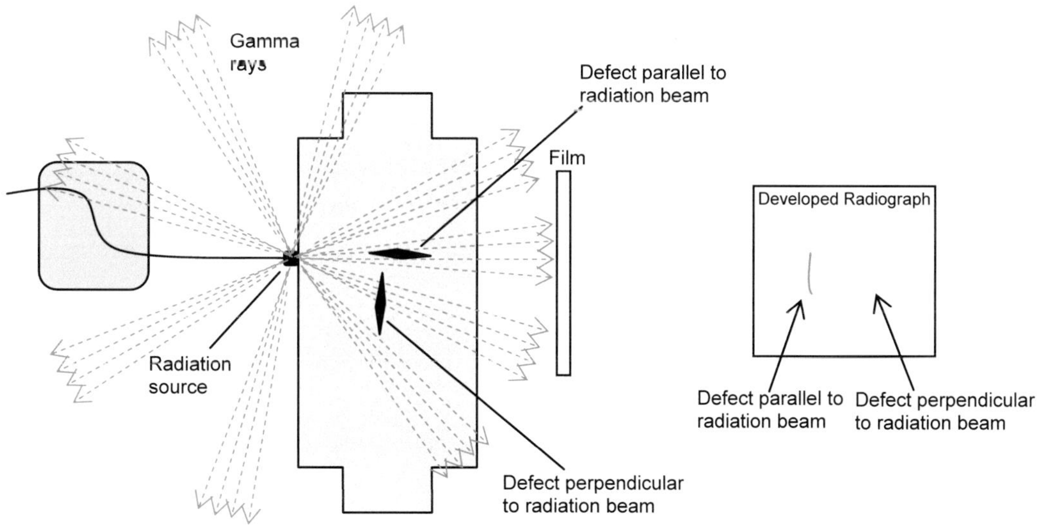

FIGURE 8-6 Cartoon of radiographic examination of a component with a parallel defect. SOURCE: Image provided by committee.

Ultrasonic testing dates back to the work of S. Y. Sokolov, a Russian scientist, in 1929. Sokolov conceived of using high-frequency, highly directional sound waves propagated through a component to detect the presence of defects by their interaction with the sound waves. Pulses of ultrasound are generated by a piezoelectric transducer which is placed on the component. The transmitted and reflected waves propagating through a component are detected with one or more sensors also attached to the component. By placing the transmitter and detector(s) at different locations, a series of images can be recorded. Features and discontinuities that are in poor contrast in one image can be observed in another image. In this respect, the technique is identical to the formation of medical sonograms used to visualize the developing fetus in utero, among many other applications. The ultrasonic images of components are generally more complex than medical sonograms because of elastic scattering (bouncing or reflection of sound waves off interfaces with significant density differences). However, depending on the techniques used, the output from ultrasonic sensors may not be a direct visual representation of the volume of the component, and so may be less readily interpreted than projection images such as radiographs (see Figure 8-7). Consequently, ultrasonic testing requires more skilled, specially trained personnel to interpret its results.

There are several characteristics of ultrasonic waves that can be utilized in addition to simple absorption used in radiography, including frequency, polarization, and phase. Major breakthroughs in the development of ultrasonic testing have included transducers that generate only shear waves, enabling other types of discontinuities to be detected; the development of time-of-flight diffraction which allows the top and bottom edges of discontinuities to be displayed and provides better accuracy in through-thickness measurement; and phased-array testing, which produces images of internal structures similar to that of medical ultrasound.

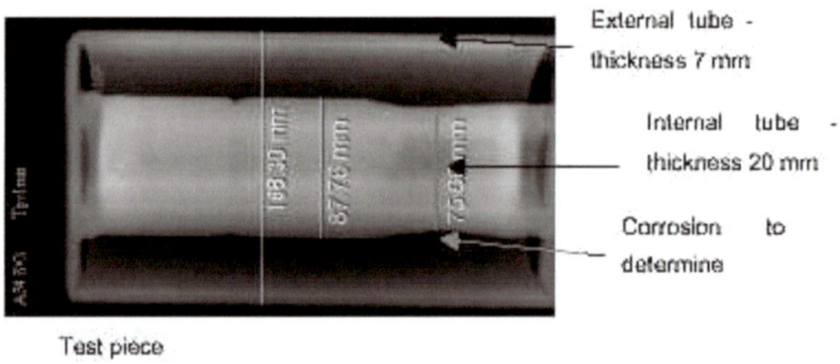

(a)

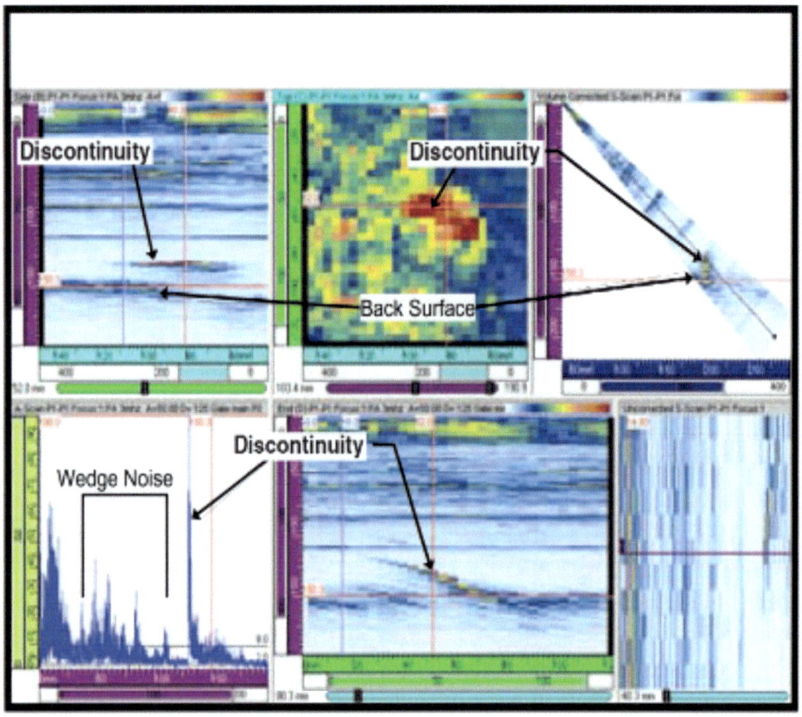

(b)

FIGURE 8-7 (a) Radiograph of a metal tube inside another larger metal tube. (b) Typical output from ultrasonic testing (of a different component) illustrating six panels of information obtained from the same component. SOURCE: (a) Blettner, A., et al. (2000), (b) Creech, M. (2006).

Over the past two decades, these and other advances in computational imaging and electromagnetic acoustic transducers have made ultrasonic techniques a viable alternative for many weld and structural inspection purposes. Already they are competing with and replacing radiography in many applications, especially in the manufacturing and quality control sectors. For instance, ultrasonic techniques are used along with radiography for inspection during manufacture of large steel pipelines used in the petroleum industry. Ultrasonic techniques are used for the inspection of critical components such as castings that are much too thick for radiography. Indeed, large metal castings or forgings, such as gas turbine rotors, can presently be examined only by ultrasonic techniques. There is also an increasing trend toward the use of ultrasonics for in-service inspection. In the railroad industry, for example, ultrasonic techniques are preferred because they are less expensive than radiography. There are also many cases in which in-service inspection by radiography is impossible, because it requires that the radiographic plate (film or detector) be placed inside an operating component. In many of these applications, ultrasonic techniques are used instead.

The costs of devices that can replace current gamma radiography in some applications vary a great deal, reflecting differences in the specifics of the inspection system, pricing for the customer, and how the system is used. Some suppliers are reluctant to offer prices for quotation in a public report, and so the figures listed here are anecdotal prices cited by customers. Portable pulsed x-ray radiography systems begin at approximately $50,000 and go up in price from there. Moveable accelerator-based radiography systems begin more in the range of $200,000. Ultrasonic systems typically range from $50,000 to $100,000. All of these cost more in maintenance and operation than gamma radiography systems.

FINDINGS

There are important and high-value applications where the simplicity, ease of inspection, and portability still favor radiography or where radiography cannot easily be replaced. These include the inspection of insulation-wrapped pipes at high temperatures, where contact methods such as ultrasonics are either not feasible or not viable economically because it would entail stripping the wrapping and cooling the pipes. Another example is in the underwater inspection of oil and gas pipelines where remote operation is essential and electrical power severely limited. Similar considerations also apply to the inspection of oil and gas pipelines in unshielded and remote locations on land.

Nevertheless, by some estimates provided to the committee (Creech, 2006), about 50 percent of industrial radiography performed today could be performed by ultrasonic methods, and a further 25 percent could be inspected using x-ray radiography as opposed to gamma radiography. With the prospect of continuing advances in computational resources, low-cost sophisticated analysis software, and image-matching algorithms, it is likely that ultrasonics and x-ray radiography (as well as other inspection technologies not discussed in this report) will further reduce the fraction of component inspections by gamma radiography. Phased-array detector technology and the other technologies mentioned are being developed for other applications. However, despite these trends, the rate at which gamma radiography is replaced may well be paced by the present, limited availability of personnel skilled in other methods, especially ultrasonics, as well as regulatory factors discussed above.

CHAPTER 9

WELL LOGGING

SUMMARY

Nuclear and nonnuclear well logging tools are used in concert with each other to obtain information about the geologic media through which a borehole has been drilled. There are five main nuclear well logging tools: the density porosity tool, using cesium-137; the neutron porosity and elemental analysis tools, typically using americium-beryllium (Am-Be) radioisotope sources; and the neutron absorption and carbon/oxygen (C/O) tools, which use 14.1-MeV neutrons from deuterium-tritium (D-T) accelerators. The cesium-137 source in a density log is a vitrified Category 3 source. The Am-Be sources range from Category 3 to Category 2. It would not be difficult to replace the Am-Be source in elemental analysis logs with a D-T accelerator. Replacing the Am-Be porosity tools is more difficult, although Schlumberger does now market D-T accelerator nuclear porosity logging tools for logging while drilling (LWD) and logging drilled holes (wireline). A major reason why such logs have not been adopted widely is that well log analysis relies on a large body of data that has been accumulated for the porosity logs using Am-Be sources. These data would be less useful in analyzing the results from 14.1-MeV neutrons from D-T accelerators. Californium-252 sources might also be used to replace Am-Be sources, but they also suffer (somewhat less) from a similar lack of supporting data. In addition, californium-252 sources have a half-life of only about 2.45 years and would have to be replaced in about two half-lives. These replacement source approaches are presently being studied by Monte Carlo simulation and in-hole experiments and demonstrations.

INTRODUCTION TO WELL LOGGING

Well logging is the practice of measuring the properties of the geologic strata through which a well has been or is being drilled. A well log is the trace or record of the data from a down-hole sensor tool plotted versus well depth. Its most common application is by the oil and gas industries which seek out recoverable hydrocarbon zones. For oil and gas production, companies would like to have several kinds of information about a geologic layer, such as the hydrocarbon content. To measure these properties, sources and sensors loaded into housings called sondes can be lowered into an existing borehole (a technique called wireline logging) or can be mounted in a collar behind the drilling head for taking measurements while the well is being drilled (called LWD).

In wireline logging, sondes and supporting electronic cartridges are strung together and lowered into an uncased borehole on a cable that has an electronic signal wire. As the string is raised, the sensors measure some or all of the following properties as functions of the depth: electrical resistivity, electron density, sound velocity, neutron moderation, thermal-neutron absorption, natural and artificial (induced) radioactivity, gamma-ray spectra, Compton scattering, borehole dimension, and occasionally nuclear magnetic resonance. The data are transmitted through the wire to computers at the surface where the data are logged.

Similar measurements can be made in a cased borehole, although it is much more challenging to carry out the measurements through the steel casing. Even with that difficulty, however, there is an increasing demand for logging of previously utilized production wells because initial hydrocarbon (oil and gas) recovery fractions may have been quite low (20–30 percent of the resources in the formation), and improved logging techniques may enable drillers to double that recovery fraction. It is much less expensive to reexamine existing wells, of which the oil companies have tens of thousands, than to drill new exploratory wells, 1,000 or 2,000 of which may be drilled each year in the southern United States and Gulf of Mexico. Small, independent well logging companies find a substantial market in relogging old wells, while the major oil field services companies (Baker-Hughes, Halliburton, Schlumberger, and Weatherford) win most of the contracts for logging while drilling.

All well logging was done by wireline until the 1980s when measurement-while-drilling and logging-while-drilling tools first became available. Measurement while drilling provides data on the location and direction of the drill head and logging while drilling gathers information about the features of a formation to the surface while the drill head is still in the formation. Crude signals can be used to carry some minimal information from the logs to the surface as the borehole is being drilled. When coupled with the ability to direct the drill head (geosteer) toward promising targets, these techniques offer several advantages over wireline logging.

With geosteering, the pitch of a borehole can be increased to a high angle relative to a vertical line to create a so-called deviated well. Such wells can follow one of the long dimensions of an oil or gas deposit, enabling much higher recovery fractions relative to a well that just traverses the short dimension of the deposit. Because of difficulties in data transmission, only critical data are transmitted, at a very low data rate (about 10 bits per second) from logging while drilling tools to the surface by mechanically vibrating the drilling fluid, which is called mud pulsing. More detailed data are stored in electronic data chips which are extracted from the logging collar when it is removed from the borehole. Another advantage of logging while drilling is in offshore drilling where the well is cased as the hole is drilled to prevent fluid intrusion into the well. In both deviated wells, where strings of instruments would get stuck, and offshore wells, where wireline logging is only possible through the borehole casing, logging while drilling is the most attractive option.

In both wireline and logging while drilling, time is a critical factor. The cost of running operations on an offshore drilling rig is very high: drilling a well might cost $1–2 million per day of operations. In such operations, downtime and logging-equipment failures are expensive. Well logging equipment costs are only a small part of the cost of drilling operations and generally a very small fraction of the hydrocarbon production costs. Modifications that improve the accuracy of logging without compromising reliability of the data are welcome in the industry even if they raise the cost. As a result, many techniques have been used for well logging. Several techniques are discussed below.

Well loggers use combinations of both radiation-based and non-radiation-based tools (called nuclear and nonnuclear tools in this field) to examine the earth formations surrounding the well and sensors to detect the media's response to interrogation tools. An analyst examines detector logs to look for some or all of the following parameters of the formation: formation water saturation, porosity, rock characteristics, carbon/oxygen ratio, and permeability.

Because of the complexity of earth formations, only a combination of all the logs allows the log analyst to draw accurate conclusions for the formation parameters. For example, combining resistivity and nuclear logs, the log analyst can determine porosity, water content, and density. Figure 9-1 illustrates a typical set of well logs. Even with multiple logs, well log analysis is an interpretive science in that it relies on data that do not uniquely determine the solution. Different well log analysts may, and often do, interpret the same logs differently. As a result, there are differences of opinion on which tools are the most important ones and which ones are valuable in what media and in combination with what other tools (see, e.g., differing views in Jacobsen et al., 2006; Badruzzaman, 2002; Ayan et al., 1999; Chang et al., 1993; Ellis, 1987).

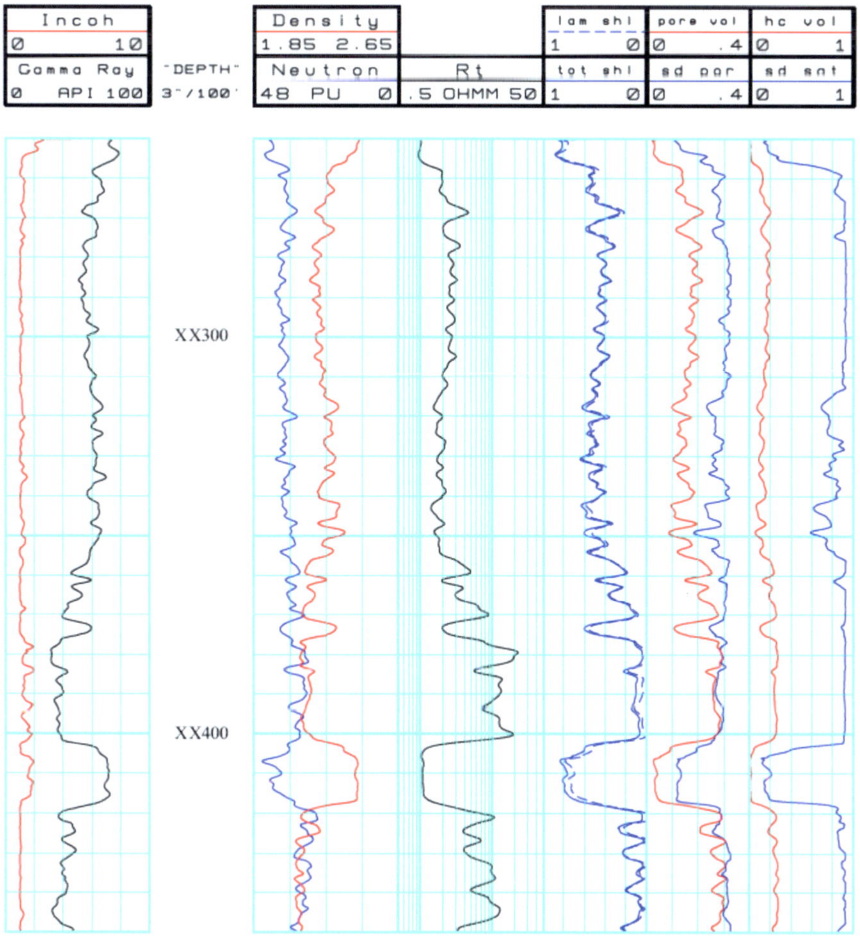

FIGURE 9-1 Typical logs and evaluation results from a string of well logging tools, including a natural gamma-ray log, a neutron log, an array induction log, and resistivity logs. The hydrocarbon volume result is also shown at the right. SOURCE: van Popta et al. (2004), courtesy of the Society of Petrophysicists and Well Log Analysts.

For these reasons, the committee has not attempted to prioritize among the different tools, and no prioritization should be inferred from the order of presentation.

The major tools for well logging are listed in Table 9-1. The tools are described in the sections that follow, first the nonnuclear well logging tools and then the nuclear well logging tools.

NONNUCLEAR WELL LOGGING METHODS

There are many nonnuclear well logging tools, including acoustic arrays, various resistivity tools, nuclear magnetic resonance tools, and formation pressure logs. Electric logs, which measure the resistivity of formations, were the first well logging technique to be used. In its simplest form, electric logging simply measures the resistivity between electrodes or transmitter/receiver coils on the sonde, which enables a well logger to identify the kind of rock within a layer and, to some extent, the likely porosity and fluid content of the pores. Most resistivity tools use induction coils through which a 20-kHz to 2-MHz signal is transmitted into the formation and induced signals from the formation are received to obtain formation resistivity.

TABLE 9-1 Well Logging Tools

Tool	Typical Method of Logging	Source of Signal	Geologic Parameter Sought
		Nonnuclear Tools	
Resistivity	LWD or wireline	Electrodes	Formation water saturation
Spontaneous potential	Wireline	Electrodes	Formation composition and water content
Induction	LWD or wireline	Coils	Formation composition and water content
Radiofrequency (rf) dielectric	Wireline	rf antenna	Formation composition and water content
Formation pressure	Wireline	Pressure in formation	Fluid content
Acoustic	Wireline	Sonic transceiver	Fluid content and porosity
Nuclear magnetic resonance	Wireline	Media in magnetic field	Fluid content and porosity
		Nuclear Tools	
Natural gamma	LWD or Wireline	K-40, Th, and U in formation	Formation composition
Gamma-gamma density	LWD or wireline	Cs-137	Formation density and shale content
Neutron cross section	LWD or wireline	D-T accelerator	Rock density and porosity
Elemental composition	LWD or wireline	Am-Be or D-T accelerator	Formation composition
Neutron moderation	LWD or wireline	Am-Be or D-T accelerator	Hydrogen content
C/O ratio	Wireline	D-T accelerator	Hydrocarbon and water content

NOTE: Am-Be = americium-241–beryllium; C/O ratio = carbon-to-oxygen ratio; Cs-137 = cesium-137; D-T = deuterium-tritium; K-40 = potassium-40; LWD = logging while drilling; Th = thorium; U = uranium. SOURCE: Provided by committee.

Occasionally, a dielectric constant tool, which measures the electrical permeability of the formation materials, is also used to help in identifying formation water content and rock types. The dielectric constant, which is another electrical characteristic of a material, is largely determined by water content in the material. The dielectric tool operates at microwave frequencies in the range of a few megahertz to 1.1 GHz.

Sonic or acoustic logging is accomplished by measuring the sound velocity in the formation from a transmitter to a sensor in the sonde. Sound velocity is an indicator of porosity and fluid content for a given type of rock. More sophisticated information can be gathered from acoustic arrays.

Nuclear magnetic resonance, which is commonly used as a medical diagnostic called magnetic resonance imaging, relies on the nuclear magnetic moment of atoms (and molecules) in a strong magnetic field and the signal they emit when they return (relax) to their original state. The technique is effective in measuring properties related to fluids, including the saturated porosity of the formation and the size of the pore spaces containing fluids, which in turn provide information about the permeability of the formation. Because of the dropoff in magnetic field amplitude over distance, the attenuation of fields in geologic media, and the mud on the walls of the borehole, nuclear magnetic resonance gathers data only on media in very close proximity to the well bore.

Direct formation pressure measurements are used to calibrate other measurements from which pressure can be inferred. The calibration enables cross comparisons of results directly and indirectly linked to the formation pressure.

NUCLEAR WELL LOGGING METHODS

There are a number of nuclear well logging tools that have been and still are important in the evaluation of hydrocarbon wells and reservoirs. While the recent interest in logging-while-drilling tools has changed the emphasis somewhat, interest in nuclear tools has remained as high as, or higher than, ever.

The nuclear tools play roles in the determination of a number of the most important hydrocarbon well characteristics such as porosity, elemental composition, and whether or not oil or water is present. The nuclear tools of primary interest use either sources of gamma rays or neutrons. The one exception to this is the natural gamma-ray tool, which has no source and detects the natural gamma rays that are present in the rock formation outside the borehole. This tool primarily identifies the depth or distance along the borehole where shale layers exist that contain naturally occurring potassium-40 or radionuclides in the uranium and thorium decay chains. The single gamma-ray tool uses a relatively small source of cesium-137 (55 to 148 GBq [1.5 to 2 Ci]), which makes these Category 3 sources) that uses gamma-ray backscatter to infer formation density outside of the borehole.

The gamma-ray source logging tool is an important tool. The source has the same measurement advantages as all radionuclide sources that have a long half-life and a known, well-defined gamma-ray energy. These advantages are: (1) they are very stable; (2) they do not require power supplies, which are often unstable and usually bulky; (3) they require very little space; (4) their radiation is monoenergetic and at an optimum energy for this application; (5) they do not require complex operational procedures; (6) they are relatively inexpensive; and (7) they emit radiation isotropically. To offset these advantages the disadvantages are: (1) they cannot be "turned off," (2) the gamma-ray energy cannot be changed, and (3) they represent a potential radiation safety risk if somehow lost or misused. These gamma sources are metal capsules containing the radioactive material, most commonly cesium-137, but other radionuclides in some cases. As noted above, these sources have relatively low activity (IAEA Category 3). They are fabricated in a vitrified form because of the aggressive environment in the boreholes. All other common nuclear logging tools use neutron sources. Two types of neutron sources are used: radionuclide sources[1] and accelerator sources. Almost all of the radionuclide neutron sources are sealed sources that contain an alpha-emitting radionuclide mixed with beryllium or boron powder that is pressed and doubly encased in stainless steel. A beryllium or boron nucleus will absorb an alpha particle and emit a neutron with energy ranging from 0 to about 11 MeV with the average energy at about 4 MeV. The most commonly used radionuclide neutron sources are Category 2 or 3 Am-Be sources, although some plutonium-beryllium (Pu-Be) sources have been used in the past. A spontaneous fission source using californium-252 has been demonstrated as a replacement for Am-Be sources. This is discussed later in this chapter.

Accelerator neutron sources are described in Chapter 4. They use an accelerated beam of ions to cause D-T fusion reactions in a target and typically produce 14.1-MeV neutrons, although with different targets they can be designed to cause different reactions that produce 2.45-MeV neutrons with much lower output.

There are four main neutron source logging tools:

1. the neutron moderation tool, which primarily measures hydrogen content that can be related indirectly to porosity;
2. the C/O tool, which measures the ratio of carbon to oxygen, which can be indirectly related to oil and/or water content;
3. the formation neutron cross-section tool, which measures the neutron absorption cross section in both the borehole and the formation outside the borehole; and
4. the mineral or elemental analysis log, which measures parameters from which the chemical and mineral composition can be inferred.

Radionuclide neutron sources are typically used for measurements of porosity and elemental analysis logs. The carbon to oxygen ratio (C/O) logs and neutron absorption logs (*n*-gamma logs) today typically use 14.1-MeV neutrons from D-T accelerators.

Figure 9-2 shows a neutron logging tool, including the relationship and dimensions of the overall well logging tool to the much smaller Am-Be neutron source. Note that the well logging tool measures over 16 meters long and incorporates a complex set of detectors and electronics. Figure 9-3 is a photograph of a gamma-gamma density instrument, showing the source-handling device used to insert and remove the cesium-137 source.

ALTERNATIVE TECHNOLOGIES

X-ray generators offer one possible alternative to radionuclide radiation sources in well logging. Some work on this topic was reported in 1986 and 1987 by engineers at Schlumberger (Becker et al., 1987; King et al., 1987; Boyce et al., 1986). They investigated the use of a commercial radiography electron linac accelerator for well logging. Although such a source of x-rays technically would provide an alternative to gamma-ray radioisotope sources, the committee judges that they were then and still are impractical due to the large size of the device. One might hope that the size of these devices can and will eventually be reduced to make this a practical alternative to the use of cesium-137 sources. However, no further development work has been reported along these lines in the intervening 20 years. In addition, even if the size limitation could be ameliorated, there are a number of other disadvantages to this approach, including the fact that the source energy spectrum is continuous, not monoenergetic; and the source studied was unstable (King et al., 1987) and had to be normalized. The source emission

[1] In well logging, these are commonly called "chemical" sources, but the committee here uses the term radionuclide sources for consistency.

is forward scatterered, not isotropic, which could be an advantage, but would need to be factored into analysis of the result.

As was already noted, there are at least four neutron source logging tools that are important in oil well logging: (1) the neutron moderation tool, (2) the mineral or elemental analysis tool, (3) the neutron cross-section tool, and (4) the C/O tool. The neutron moderation tool and the mineral or elemental analysis tool typically use the radionuclide neutron sources because these tools are more effective if the initial neutron energy is lower. Neutrons emitted by Am-Be sources have energies substantially lower than D-T fusion neutrons (average energies around 4 MeV versus 14.1 MeV). The C/O tool and the neutron cross-section tool normally use an accelerator D-T source because these tools use the gamma rays produced by neutron inelastic scattering, which favors higher neutron energies and is more easily measured if the neutron source is pulsed.

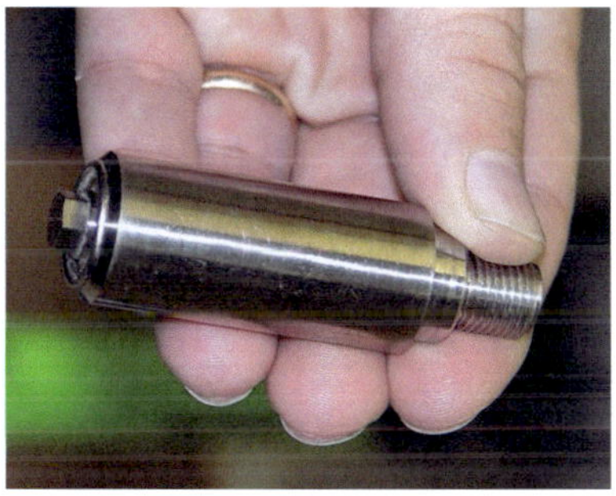

a)

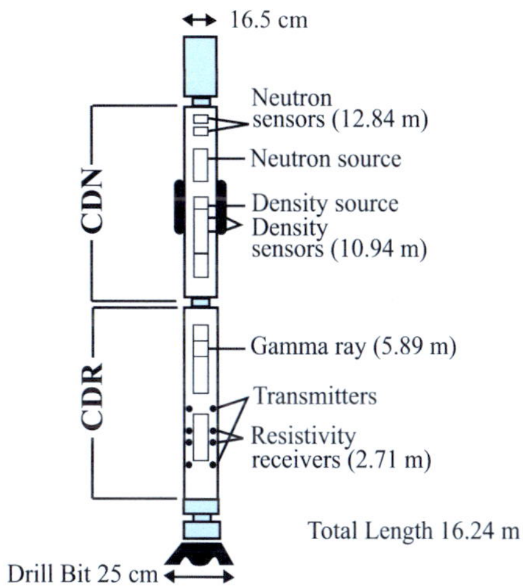

b)

FIGURE 9-2 Photograph of an Am-Be neutron source used in well logging (a) and a diagram of a sonde mounted on a drill for logging while drilling (b). SOURCE: (a) Image courtesy of Sandia National Laboratory (William Rhodes, III), (b) Provided by the committee.

FIGURE 9-3 Atlas Densilog, which uses a vitrified cesium-137 source for wireline density measurements. The rod extending downward is a source handling device. The light-colored circles are windows for the source and detectors. SOURCE: Photo courtesy of Baker Hughes Inc.

The neutron moderation tool measures the slowing down of neutrons by detecting the radiation (gamma or neutron) scattered back to the detector as the neutrons undergo collisions in the media. Higher energy neutrons penetrate more deeply into the media before much of the radiation is scattered back, and the intervening media shields the radiation to some extent. The high-energy neutrons emitted by the source are slowed to about thermal energy by the multiple scattering of the neutrons, primarily with hydrogen atoms, before they are detected by detectors that have their highest probability of detection with thermal energy neutrons. On average this may require anywhere from 15 to 200 scattering interactions per neutron from an Am-Be source (depending on the hydrogen content of the medium) and the average neutron path may be on the order of 50 cm. When 14.1-MeV neutrons are used, more scatters are required and the resulting average path length is longer, if only elastic scattering is of importance. As oil reserves are tapped (and formation saturation decreases) the neutron path length grows longer. To detect the same number of thermalized neutrons for the 14.1-MeV case, one requires a more intense source or a more sensitive detector: Because the neutrons penetrate more deeply and scatter farther away, the neutron detector would have to be placed at a greater distance from the source and the intensity of the signal diminishes as the inverse of the square of the distance. The same type of reasoning also applies to the elemental analysis tool except that in this case a longer distance would be required for neutron moderation before the (n,γ) (pronounced "n-gamma") reaction occurs, which then produces a gamma ray that returns to the detector in the logging tool. So in both cases, a more intense source of neutrons or an improved detector would be required for the 14.1-MeV neutrons and a larger sample would be interrogated.

However, elastic scattering is not the only mechanism of importance in slowing neutrons in well logging. Whenever oxygen and carbon are present (which is usually the case), inelastic scattering of high-energy neutrons becomes more important and neutron elastic scattering becomes less important. In this case, fewer elastic scatters may be required to thermalize neutrons for high-energy-source neutrons and the source-to-detector spacing required may be less for a higher energy neutron source. It is clear that this logging principle is complex.

The elemental analysis tool relies on the (n,γ) reaction in the geologic media, and that reaction is more probable at lower energies. In some materials, thermal neutron energies are best.[2]

Present accelerator sources are sealed and have a lifetime of several hundred hours; field service companies often bring two sources to a job in case one fails. Changing the sealed-source part of the accelerator is an added expense for oil-well logging companies, but this expense is justified because the pulsed, high-energy neutron source is advantageous. Note that there are some liabilities associated with accelerator sources, including the need to keep two tools on hand at any job site in case of tool failure and the regulatory and logistical burden caused by the dual-use nature of the D-T neutron generators (see Sidebar 9-1). Also, as mentioned in Chapter 4, D-T neutron generators do not obviate the need for radioactive waste management, although they constitute a lesser burden than the radionuclide radiation sources.

[2] Thermal neutrons have energies corresponding to the temperature of the medium through which they travel. A typical average energy for thermal neutrons is around 0.025 eV. The average energy of neutrons from an Am-Be source is nearly 200 million times higher.

SIDEBAR 9-1
Sealed D-T Neutron Generator Systems Are Nuclear-Related Dual-Use Technology

Dual-use technologies are those that have both commercial and military or proliferation applications. Because a sealed pulsed neutron generator (one with no external vacuum system) can be used as a trigger for a nuclear explosive device, it is dual-use nuclear equipment and subject to the Export Administration Regulations, administered by the Department of Commerce. Federal regulations impose requirements on both domestic and international shipments and use of these accelerator-based neutron sources. Domestic shipments must be tracked. The regulations require that International shipments have an export license for each movement of a device outside the country, restrict the number of such shipments to and from each country, and impose both administrative costs and often delays in shipment. The regulations are the U.S. mechanism for implementing an international system of nonproliferation measures called the export control regime.

One service company, Schlumberger, markets accelerator neutron porosity tools: a wireline tool called APS and an LWD tool called Ecoscope. APS was introduced in the early 1990s and has gained modest acceptance. It relies on detection of epithermal neutrons, and so the results are not readily comparable with results found with other tools. Ecoscope was brought to the field in 2005 and utilizes a different detector array which, with analysis, provides results that are more directly comparable to those from other tools (H. Evans, QSA Global, Verbal presentation to the Committee on Radiation Source Use and Replacement, Irvine, CA, February 1, 2007). Note that the neutron flux output from these tools is reported to be higher by a factor of 5 to 10 than that from Am-Be sources in their radionuclide-source counterpart tools. Schlumberger declined to quote the price of its tools for this report. Further, the company sells services using the tools, not the tools themselves. Anecdotal reports of costs for D-T tools that are for sale by other companies range between $40,000 and $50,000, which provides a rough estimate of the cost of the D-T porosity tools, although this range does not account for differences in the detector costs and the supporting software.

Another interesting possibility is the replacement of Am-Be with californium-252 spontaneous fission sources. Investigations into using californium-252 for well logging date back to the late 1960s, but to the committee's knowledge, only Pathfinder Energy Services provides such a tool for LWD (see Valant-Spaight et al., 2006). Californium-252 sources (described in Chapter 2) emit fission neutrons and gamma rays with a neutron energy spectrum skewed lower (average neutron energy of about 2.14 MeV) than the Am-Be sources (average neutron energy of about 4.18 MeV). The Pathfinder tool uses a 0.41-GBq (0.011-Ci) californium-252 neutron source, which is a Category 4 source, instead of a 300-GBq (8-Ci) Am-Be source to emit neutrons at the same rate (1.8×10^7 neutrons/s) after 3.5 years. Valant-Spaight et al. (2006) report that the californium-252 source emits a factor of 15 lower gamma-ray exposure rate than the equivalent Am-Be source (although this may refer to the equivalent Am-Be source at the initial neutron emission rate, which is 740 GBq or 20 Ci). The californium-252 source costs approximately $6,000, and the tool costs in the range of $100,000 (Mitchell Ferren, Oak Ridge National Laboratory, personal communication with M. D. Lowenthal, June 13, 2007; W. Schultz, Pathfinder Energy Services, personal communication to M. D. Lowenthal, June 21, 2007).

Both the californium-252 sources and the D-T accelerator sources have shorter working lives than the Am-Be sources. The californium-252 sources have a half-life of 2.65 years and need replacement after 4 or 5 years (they begin with a larger neutron output to enable longer operation). The operating life of pulsed neutron generators varies but typically might be around 500 hours. The Am-Be source, by contrast, has a half-life of about 433 years, and so it only needs to be replaced when it requires reencapsulation, which the manufacturers recommend as every 15 years.

There is some incentive for well loggers to switch away from Am-Be sources. After the U.S. domestic supply from the Department of Energy's (DOE's) Isotopes Program was exhausted, REVISS was left as the only international supplier of americium sources, and the price of a well logging source climbed dramatically. Anecdotal reports of the costs are in the range of $80,000 to $100,000 each, with a two-year lead time on orders. There is a stock of Am-Be sources at Los Alamos National Laboratory collected through the Offsite Source Recovery Project. Some in the well logging industry have sought to have DOE recycle these sources, making them available to well loggers. The sources are planned to be disposed and no decision has been made within DOE to recycle them. In addition, the new U.S. NRC requirements for handling Category 2 sources has caused several companies to consider redesigning their tools to use sources just below the Category 2 threshold (0.6 TBq or 16 Ci for Am-Be).

Even with the high cost of Am-Be sources, both the D-T accelerator neutron source and the californium-252 neutron source face a significant obstacle to being adopted more broadly within the well logging industry: The analyses of well logging data rely on a large body of data that has been accumulated for the porosity logs using Am-Be sources. These data would be less useful in analyzing the results from 14.1-MeV neutrons from D-T accelerators and fission neutrons from californium-252 sources until a substantial database using these sources accumulates. Even if these sources hold the promise of providing superior information, this database inertia must be overcome. Both D-T and californium-252 source replacements are now being studied by Monte Carlo simulation and, to some extent, by in-hole experiments, demonstrations, and operations with tools on the market, but more will be needed to make these replacements more broadly attractive. Schlumberger would argue that its Ecoscope tool is already being accepted. Others point out that the well log databases are not static: No two geological fields are the same; conditions change even in the same field; the same genre of tools from different vendors can give significantly different interpretations of petrophysical parameters (Badruzzaman, 2002); and new technologies are adopted because they add valuable information, and any new tool, even with the same old radionuclide sources, requires extensive calibration before (and after) it is fielded. Even accepting all these points, however, the committee observes a barrier to acceptance of a replacement tool that gives a measurement that differs from the old one. Field service companies report that upon introduction of a new tool, their customers ask if the tool can provide a result that looks like the old measurement.

There is not an obvious role for the federal government in overcoming this and other obstacles to implementation of these alternatives. The industry itself, however, can form industry working groups, called Special Interest Groups under the Society of Petrophysicists and Well Log Analysts, to investigate questions and establish common practices across the industry. Such a group already exists concerning nuclear well logging tools. If so tasked, the group could develop new reference standards (measure standard signals from known reference rock formations) for these replacement tools, examine the response of these tools relative to the Am-Be tools, and explore any differences in response when the replacement tools are used in combination with other nuclear and nonnuclear well logging tools.

Detector Technology

The well logging industry continuously monitors developments in detector technology. Although this technology is being researched heavily by a rather large industry, advances appear to be relatively slow, especially for those that would drastically affect hydrocarbon well logging. Well logging for hydrocarbon recovery requires very rugged detectors that can be exposed to and will operate in the harsh environment of the well borehole. As noted in Chapter 4, so far the primary detectors being used are the sodium-iodide (NaI) scintillation detector for the natural gamma-ray and density backscatter gamma-ray tools and the helium-3 (He-3) gas proportional counters for neutron detection. The latter (He-3) detector replaced the gas proportional BF_3 detectors perhaps 10 to 15 years ago. There is some recent use of bismuth germinate (BGO) scintillation detectors in place of the NaI detectors, but the trade-offs apparently do not greatly favor this replacement at present. Although the BGO detector has higher density and effective atomic number, which gives better total and full-energy peak detection capability, the temperature response and scintillation efficiency characteristics still favor NaI detectors. However, BGO is not the only higher density scintillation detector that is used as an alternative to NaI. A gadolinium silicate crystal (Gd_2SiO_5 or GSO) is available, which is slightly less dense than BGO but has better temperature characteristics than BGO (Roscoe et al., 1991). For more detail on detector technology, one should refer to Chapter 4 of this report, the standard text by Knoll (2000), and articles in the journal *Nuclear Instruments and Methods,* Parts A and B.

FINDING AND RECOMMENDATION

Finding: Accelerator neutron sources and californium-252 sources show promise as potential replacements for americium-beryllium sources in neutron well logging tools. However, there are technical obstacles for these replacement sources and they are at a disadvantage based on the extensive experience and data accumulated with americium-beryllium sources.

Recommendation: The Society of Petrophysicists and Well Log Analysts should task an industry working group, called a Special Interest Group (SIG) such as the Nuclear Logging SIG, to address the technical obstacles to implementing replacements for the americium-beryllium sources used in well logging and the challenges of data interpretation. The group should decide what obstacles are most important, but the issues might include development of new reference standards for these replacement tools, examination of the response of these tools relative to the americium-beryllium tools, and exploration of any differences in response when the replacement tools are used in combination with other nuclear and nonnuclear well logging tools.

Replacement of the cesium 137 sources used in well logging would be difficult because well loggers desire monoenergetic gamma rays, compact sources, and robust source forms for the cesium sources. These sources are also vitrified and not easily dispersed and are of Category 3 intensity. Thus, the committee judges that replacement of these sources is not a priority.

CHAPTER 10

IMPLEMENTATION OPTIONS FOR ENCOURAGING REPLACEMENT OF RADIONUCLIDE RADIATION SOURCES WITH ALTERNATIVES

Replacement technologies for Category 1 and 2 radionuclide sources are available or possible for nearly all irradiation applications. The existence of commercial x-ray and electron-beam sources for some applications indicates that in some circumstances the marketplace has assessed the currently available alternative technologies to be financially feasible. The full social cost of radionuclide use includes some costs not borne by the users, including the costs of safe disposal for some radiation sources (a component of the other social costs (OSC), introduced in Chapter 1) and the costs associated with the risks that the radionuclide radiation sources might be used in acts of terrorism (TRC, introduced in Chapter 1). The extent to which the user should be responsible for costs of disposal and the costs that might be incurred in association with radiological terrorism deserves some debate.

The availability of alternative technologies and a desire to reduce the total social costs of radionuclide use motivate the committee's recommendations that the U.S. government take steps to promote the replacement of high-risk radionuclide sources with lower risk alternatives. As noted in Chapter 3, the committee's charge directs the committee to make findings and recommendations on options for implementing the identified replacements," and the U.S. Nuclear Regulatory Commission (U.S. NRC) asked that these include both technical and policy options. The committee considered a range of policy options that could be employed to encourage or implement replacements. The U.S. NRC informed the committee that the agency examines proposed uses of radioactive material to evaluate whether they protect public health and safety and promote common defense and security; not whether the uses conform to policy goals beyond those considerations.[1] The committee did not consider the specific legal or regulatory authority required for implementation of policy options, whether the policy options fit within the common defense and security clause of the U.S. NRC mission, or if new legislation would be needed to enable the U.S. NRC or another agency to carry them out, and the committee makes no claims about these questions.

An array of different policies could be adopted to promote radionuclide replacement. The sections that follow first set out a generic menu of policy options to assist policy makers in considering the range of policies available, and then discuss particular applications to address the risks posed by devices using radioactive cesium chloride. The same policy options could be applied to replacement of other radionuclides, americium-241 sources being the most similar to radioactive cesium chloride both with respect to hazard and with respect to the lack of disposal options. The committee does not here emphasize these other radiation sources because (1) the committee considers radioactive cesium chloride to be the top priority; (2) there are already incentives in the fields of radiotherapy, radiography, and well logging that make professionals in those fields seek nonradionuclide alternatives. In radiotherapy, teletherapy units have shifted from cobalt-60 to linear accelerator (linac) sources, and companies with linac products are trying to compete with the Gamma Knife®, which is the only Category 1 or 2 radiotherapy device that

[1] The U.S. NRC's mission, as described in its strategic plan (U.S. NRC, 2004) is to "License and regulate the Nation's civilian use of byproduct, source, and special nuclear materials to ensure adequate protection of public health and safety, promote the common defense and security, and protect the environment."

is now increasing in use in the United States. In radiography, the inconvenience of imposing radiation protection procedures at job sites and the inherent limitations of radiography have driven practitioners to shift to other techniques for nondestructive inspection, such as phased-array ultrasound. In well logging, there are barriers to switching away from americium-beryllium sources, but the difficulties, costs, and exposures associated with these sources give the industry cause to look for better practices. It is clear to the committee that the large contract irradiator companies do not yet see strong incentives to shift from gamma irradiation to x-ray irradiation, and so this may be another area where additional encouragement is needed. Again, however, the committee's discussion focuses on radioactive cesium chloride.

Finally, mention of a policy option in this chapter does not constitute an endorsement of that option. Indeed, some are mentioned to highlight their undesirable qualities and consequences.

GENERIC POLICY APPROACHES

Table 10-1 summarizes four classes of generic policies: prohibitions, push incentives, pull incentives, and supply incentives. For each of these classes, several possible generic policies and their major advantages and disadvantages are shown.

Prohibitions are the most direct way to eliminate radionuclide use. They force either replacement or abandonment of use. Rescinding already-issued licenses would be extremely costly in some cases, in light of investments made by users in anticipation of the continuation of the licenses, and would require compelling arguments to support the action. Even when narrowly applied, prohibitions are very blunt in removing both uses for which replacements are readily available as well as those for which particular circumstances make replacement infeasible or extremely costly. Because of these disadvantages, rescinding already-issued licenses is likely to be neither feasible nor desirable. Some observers may argue that if the U.S. NRC determines that a set of radiation sources pose substantial risks, then it should impose a swift, categorical prohibition. Few situations, however, offer clearly unacceptable risks.[2]

Prohibitions on new licenses (i.e., no sales or import of sources) offer more promise. Although they leave the existing stocks of radionuclide sources in place, they effectively cap the total number so that over time there will be a decline as the sources decay and units are retired. The determination of which uses to no longer license requires confidence in the existence of commercially viable replacement technologies. Even if replacements are commercially viable in general, there may be specific applications for which replacement is commercially infeasible. Therefore, prohibitions on new licenses would have to be carefully targeted to avoid losing benefits of radionuclide use that cannot actually be replaced with current technology.

Push incentives seek to make replacement technologies relatively more attractive to potential adopters by internalizing more of the external costs (TRC + OSC) of use of radionuclides, as discussed in Chapters 1 and 3.

One way to accomplish this is to impose more stringent requirements on users. The most desirable requirements would reduce the risks associated with use. For example, additional requirements to ensure physical security or the quick discovery of diversions would both reduce the risks and increase the costs of radionuclide use. The increased costs would make replacement technologies relatively more attractive. Developing effective regulations,

---

[2] The selective banning of the type of sources involved in the Decatur, Georgia incident is one example. The banning of a certain design of connectors for attaching radiography source "pigtails" to drive cables is another. The latter action was taken following an accident in California in 1979 in which a plant worker received a serious radiation burn when he picked up a radiography source that had detached from the drive cable unnoticed by the radiographer (U.S. NRC, 1982).

especially in light of the diversity of uses, would entail agency costs in designing restrictions that actually reduce risks and costs in enforcing those restrictions.

TABLE 10-1 Generic Policies for Promoting Radionuclide Replacement

Generic Policy	Major Advantages	Major Disadvantages
	Prohibitions	
Complete ban — Rescind existing licenses for particular radionuclides or classes of use	Rapid and complete replacement	Can be very costly to users (or the party that pays); inflexible
New source ban — Stop issuing new licenses for particular radionuclides or classes of use	Caps number of uses	Slow; inflexible
	Push Incentives	
More stringent regulations — Require investments by users in risk reduction	Directly reduce risks; make replacements relatively more attractive	Regulatory costs; regulations may not sufficiently promote replacement
Use fees — Raise monetary cost of use through fees on particular radionuclides	Makes replacements relatively more attractive; generates revenue for other uses	Difficult to choose appropriate rate; administrative costs
Decommissioning funds — Impose full dispositioning costs at time of purchase	Internalizes disposal into technology choice	May be prohibitive in absence of disposal options; administrative costs
	Pull Incentives	
Direct subsidies — Offer payments for particular retirements or replacements	Marginally encourages flexible replacements	Budgetary cost; administrative costs
Tax subsidies — Reduce cost through allowed deductions for replacements	Marginally encourages flexible replacements	Revenue loss; not applicable to nonprofits
Buybacks — Offset scrapping costs by purchasing particular classes of devices	Encourages flexible replacements, especially for older devices	Budgetary cost; administration of physical disposal
	Supply Incentives	
Supplier subsidies — Research and development grants	Encourages improvements in replacement technologies	Administratively difficult to pick good projects; budgetary cost
Certification services — Provide publicly funded testing and certification services for replacement devices	Encourages improvements in replacement technologies; no need to pick winners	Budgetary cost

SOURCE: Provided by the committee.

A more direct action that increases the cost of radionuclide use is to impose a fee on some aspect of use such as the activity in inventory. For example, a fee could be set on the net addition of activity from radioactive cesium chloride that would be paid by sealed-source providers; a lower fee could be set on net-curie suppliers provided in less dispersible forms of cesium-137 to encourage their development. The economically efficient fee would be exactly equal to the external costs of use (the social costs not previously borne by users). However, because the external costs cannot be confidently monetized, a fee would have to be set on some other basis, such as providing a sufficiently strong incentive to encourage replacement without making use so prohibitive that specialized applications without viable replacements become commercially infeasible. An advantage of the fee approach is that it would generate revenue that could be added to general revenues or earmarked for subsidy programs to encourage replacement or run an insurance pool to handle major radiological incidents. Most regulators are empowered to recover regulatory costs through fees and to impose fines for infractions by licensees, but not to impose fees such as those discussed here (for fear that they will use the power inappropriately to raise funds).

Another action that would increase the cost of radionuclide use is to require licensees to provide larger decommissioning funds. The U.S. NRC requires licensees with inventories of radioactive material exceeding certain thresholds to set aside funds, either as segregated accounts or through financial intermediaries, to cover a portion of the estimated costs of decommissioning (see Sidebar 10-1). The quantity thresholds on sealed sources currently in place do not subject all license holders to the regulations on providing financial assurance for decommissioning. For example, there is no requirement for financial assurance for decommissioning of any self-contained irradiators with cesium-137 sources currently in use because these devices do not contain over 3,700 TBq (100,000 Ci). The thresholds could be lowered to include more licensees operating sealed sources or the provisions made more demanding to make the use of certain radonuclides more costly.[3] If disposal costs could be confidently monetized, then a decommissioning fund arrangement that required users to bear these costs in full would be desirable to internalize OSC in their private decision making.

Because production of more robust forms of cesium-137, such as pollucite, require investment in new production lines, suppliers would have to be convinced that a sufficient and sustained demand exists for the product. Actions that tend to ensure that there will be a continued demand for these new radionuclide radiation sources at a somewhat higher price would tend to lower one of the barriers to availability of this alternative.

Pull incentives encourage the adoption of replacement technologies by lowering their cost to potential adopters. Direct subsidies involve cash payments made from federal funds to adopters of specific technologies. Implementation would require a budget allocation, the determination of qualifying investments, and an administrative agency to distribute the funds. As with a user fee, determination of the efficient level of subsidy would depend on the monetization of external costs of the radionuclide source being replaced. The efficient subsidy would equal the avoided external costs, TRC + OSC. The subsidy could be provided without an explicit budget allocation through an income or profits tax deduction or credit. Neither of these tax subsidies would be relevant to not-for-profit organizations such as the American Red Cross, which uses cesium chloride blood irradiators.

---

[3] In its 2006 position statement, the Health Physics Society recommended this approach: "The HPS recommends that a requirement be incorporated into the licensing process that an acquirer of Category 1, 2, or 3 sources must provide financial surety for disposal of the sources. This financial surety could be, for example, via an escrow account under NRC control with sufficient funds to cover government or third-party costs to dispose of the sources on the license with return of remaining funds to the purchaser upon disposition of all sources and termination of the license. The establishment of financial surety is consistent with the IAEA Code of Conduct" (Health Physics Society, 2006).

SIDEBAR 10-1
Financial Assurance Requirements on Sealed-Source Licensees for Decommissioning

Some sealed-source licensees are required to provide financial assurance for decommissioning of their sources, devices, or facilities, in the form of prepayment or an external sinking fund, surety bond, a dedicated letter of credit, insurance, or some other guarantee. The U.S. NRC's requirements for financial assurance are described in regulations and guidance (10 CFR § 30.35 and NUREG-1757), and they apply to licensees with sealed sources within specified activity ranges. The ranges and the requirements are described in the table below. The funds required for licensees with quantities of material in the highest category are reevaluated every three years. Financial assurances are not required for radionuclides with half-lives less than 120 days, and so a licensee with only iridium-192 is not required to have financial assurances for decommissioning in place. A different set of activity ranges applies to unsealed sources. For unsealed sources, the limits are lower and the required funds are higher, presumably in light of the greater potential for contamination from unsealed sources.

Under a decommissioning plan, the financial assurance amount must include estimated costs for disposal. The program has not had to confront the problem of estimating disposal costs for radiation sources that would require disposal as Greater-than-Class-C waste (high-activity cesium-137 and americium-241 sources) because nearly all of the sources in the United States subject to this requirement are below the activity thresholds for requiring financial assurance for the current year, 2007.

2007 Financial Assurance Requirements for Decommissioning

Licensees with Sealed Sources Containing Byproduct Material in the Following Quantities	Financial Assurance Requirement	Examples of Radionuclide Activity Limits (based on quantities in Appendix B of 10 CFR Part 30)	Examples of Typical Sealed Sources or Devices in this Range
Less than 10^{10} times the quantity in Appendix B of 10 CFR Part 30	None	Less than 3,700 TBq (100,000 Ci) of cesium-137, OR Less than 370 TBq (10,000 Ci) of cobalt-60, OR Less than 3.7 TBq (100 Ci) of americium-241	34 cesium-137 self-shielded blood irradiators, OR 1 or 2 cobalt-60 self-shielded irradiators, OR 4 Am-Be neutron well logging sources
More than 10^{10} times but less than 10^{12} times the quantity in Appendix B of 10 CFR Part 30	$113,000	3,700 to 370,000 TBq (100,000 to 10,000,000 Ci) of cesium-137, OR 370 to 37,000 TBq (10,000 to 1,000,000 Ci) of cobalt-60, OR 3.7 to 370 TBq (100 to 10,000 Ci) of americium-241	1 or more cesium-137 panoramic irradiators OR 1 to 66 new cobalt teletherapy heads OR 6 to 500 Am-Be neutron well logging sources
More than 10^{12} times the quantity in Appendix B of 10 CFR Part 30	Site-specific decommissioning funding plan based on estimated decommissioning costs	More than 370,000 TBq (10,000,000 Ci) of cesium-137, OR 37,000 TBq (1,000,000 Ci) of cobalt-60, OR 370 TBq (10,000 Ci) of americium-241	No known applications of cesium-137 sources in this range, OR 1 cobalt-60 panoramic irradiator, OR No known americium-241 sources in this range

SOURCE: 10 CFR § 30.35

A buyback program could also be used to subsidize the adoption of replacement technologies by current users of radionuclide devices. Currently, the cost of disposing of a device may be sufficiently high that it becomes a barrier to switching technology. It could also be a barrier to disposing of devices that have already been replaced by alternative technologies. For example, the disposal costs of a cesium research irradiator may add tens of thousands of dollars to the cost of switching technologies. Even when the irradiator is no longer being used, the disposal costs may lead owners to store the device on-site, perhaps with less attention to security than would have been the case were it still in use. Further, irradiators that are disposed commercially may be recycled for use in other countries, where a diversion risk would remain, or even increase. A buyback program, which would require a federal budget allocation and appropriate physical facilities for disposal or storage, could be designed to offset the disposal costs, removing a barrier to adopting replacement technology. It would also help sweep up no longer used devices and prevent them from being reused in other countries.

Supply incentives seek to encourage device makers to develop and promote better replacement technologies. The various pull incentives could be given to firms that supply irradiation devices to end users. Their impacts would in general be similar to the pull incentives directed at users. A different pull approach that operates through suppliers is a program of research and development (R&D) grants aimed at improving replacement technologies. Consider, for example, R&D to address x-ray generation inefficiencies in the gray zone between 0.5 and 1.5 MV. These grants would lower the costs to suppliers of conducting R&D. Aside from their budgetary costs, they would also require considerable administrative oversight to select promising projects and ensure that grants are used appropriately.

Another example of useful R&D is qualification of alternative matrixes for high-activity cesium-137 sources. R&D on producing more robust matrixes for high-activity cesium-137 sources could make it easier or less costly to provide lower hazard cesium-137 sources. Such research could be most effective if carried in partnership with investigators and facilities in Russia, where experts know more about what is already done and where hot tests (experiments with actual cesium-137) are possible. A natural starting point for such R&D is to convene a small international technical meeting on matrixes for cesium-137 where invited specialists from PA Mayak and other institutions could discuss current production methods, future prospects, and R&D needs. Such a meeting could be organized by a U.S. institution or agency or through the International Atomic Energy Agency (IAEA).

If supply incentive options are attractive but policy makers seek lower administrative costs, then the federal government could offer to provide free testing and certification (to be carried out at national laboratories or by contractors) of new or modified devices that improve the risk profile. Testing and certification are required for licensing of equipment such as blood irradiators. Some device manufacturers and distributors cite the costs of redesign and recertification as barriers to bringing alternative technologies to market. In this scheme, the federal government would provide no direct grants or tax incentives and would not be put in a position of "picking winners." Businesses would still bear the cost of redesign, which is a proprietary matter, with costs based on a variety of business decisions. The more standardized cost of testing and certification of a limited number of designs could be borne by the public through the federal government.

Phase Out Use of Category 1 and 2 Cesium Chloride Devices

In Chapter 3, the committee makes findings and recommendations concerning radioactive cesium chloride sources. Here, the committee describes the steps suggested as options for implementation of the replacement of these sources, beginning by recapping the rationale for action.

Because the potential for area-denial radiological dispersal device (RDD) consequences are significant (perhaps the most significant) hazards associated with malevolent use of Category 1 and 2 sources, evaluations of security requirements for licensees should account for these consequences. The U.S. NRC, upon review, may determine whether to apply additional security requirements to some licensees. These are near-term actions.

A full assessment of hazard that goes beyond deterministic health effects might very well lead the U.S. NRC to increase the security requirements for new cesium chloride irradiators to levels that would result in few new units being licensed. Indeed, the imposition of more stringent safety requirements based on a comprehensive assessment of risk by the Canadian Nuclear Safety Commission effectively stopped the licensing of new cesium chloride irradiator designs and has resulted in an overall decline in Category 1 and Category 2 cesium chloride sources by approximately 50 percent since 2000 (R. Jammal, Canadian Nuclear Safety Commission, verbal communication to the Committee on Radiation Source Use and Replacement, January 9, 2007; see Sidebar 10-2).

SIDEBAR 10-2
Canadian Nuclear Safety Commission Actions Requiring Source Certificates

The Canadian Atomic Energy Control Board (AECB) was established under the Atomic Energy Control Act of 1946. Its mission was to supervise and control the development, application, and use of atomic energy. In 2000, the new Nuclear Safety and Control Act and its regulations came into force and the Canadian Nuclear Safety Commission (CNSC) replaced the AECB, with an accompanying overhaul of the regulatory structure and approach. Licensing of sealed sources is just one area where regulatory changes have led to changes in practices.

The AECB had "approved" devices that were not formally certified according to current standards and practices, as promoted by the IAEA and others. Under the new CNSC, certification was required for all sources and devices, which forced reexamination of every device design for certification during the three year transition period from the old system to the new one. Device manufacturers submitted for certification many of their devices because the manufacturers would have been unable to sell or service the devices that did not receive certification. Licensees with those devices supported by the manufacturer felt no additional burden except some requirements to enhance security. Some devices were no longer supported by the device manufacturers, in which case the burden of seeking certification fell upon the licensee. In addition, each device now requires a special form certificate that must be renewed at least every five years. Special attention was given to radioactive cesium chloride devices because of the hazards they pose to worker and public safety and security.

This action, along with the enhanced security requirements for sealed sources and devices, significantly reduced the number of Category 1 and 2 radioactive cesium chloride sources held by licensees. It did not, however, prohibit the sale of new devices or reloading of existing devices that were recertified and supported by a manufacturer. Many licensees with devices not supported by the manufacturer concluded that the certification was not worthwhile, particularly for underutilized devices, the services of which could be more efficiently provided by a central facility. The licensee cost of compliance was not considered by CNSC in imposing these requirements, because the requirements were considered matters of safety and security.

The government of Canada also provided a place for retired radioactive cesium chloride sources to go. For a fee, Canadian companies licensed to service (decommission, decontaminate, and dispose) radiation sources and devices can take sources and dispose of them in the Atomic Energy of Canada, Limited disposal facility. This is true even for foreign radiation sources and devices that are decommissioned by an appropriately licensed Canadian company.

The use of Category 1 and 2 cesium chloride sources involves substantial external costs because of the hazard posed by the potential for malevolent use (TRC) and the absence of an avenue for permanent disposal (OSC). Further, feasible replacement technologies currently exist or could be introduced for nearly all applications. The combination of large externalities of use (liabilities) and the availability of alternatives led the committee to recommend that the government take steps to phase out the use of these sources. The suggested options follow.

Discontinue Licensing of New Category 1 and 2 Cesium Chloride Sources

The Canadian experience suggests that more stringent regulatory requirements, based on a fuller assessment of consequences, could reduce the number of existing cesium chloride devices but not eliminate new ones of approved design. An explicit ban on new Category 1 and 2 cesium chloride devices would ensure that the number of such devices would not increase. The ban would be more effective if it applied to both new devices and the recharging of existing devices. It would signal to suppliers that there would be a market in the future for alternative technologies. As previously noted, at least one x-ray irradiator is already a commercially available alternative, and more may be on the way. In some applications, cobalt irradiators, which are less common and the committee views as less hazardous than cesium chloride irradiators, may be feasible alternatives.

An even more stringent prohibition would ban all new Category 1 and 2 cesium sources rather than just those devices employing cesium chloride. The less stringent ban, however, creates an incentive for suppliers to offer devices employing cesium in glass or mineral (pollucite) form. The committee judges that these alternative forms would substantially reduce the hazard associated with cesium chloride sources. Some form of incentive appears to be needed to encourage source producers and manufacturers to provide high-activity cesium-137 sources in these alternative forms because they are more difficult and more costly to fabricate, and there is little evidence now of a customer demand for them.

As noted above, the committee did not analyze whether the ban on radioactive cesium chloride could be implemented under existing U.S. NRC authority. Legislative authorization may be necessary.

Incentives for Decommissioning Existing Sources

The over 30-year half-life of cesium means that a cesium chloride irradiator can operate effectively for 20 to 30 years without being recharged. The profile of private costs includes a large initial capital investment, low operating costs (in the absence of stringent safety requirements), and a negative scrap value (i.e., decommissioning generally involves paying a scrapping fee rather than receiving a payment for the used device). For a cesium chloride irradiator already in operation, the capital costs are sunk in the sense of no longer being relevant to assessing the private future lifetime costs of use. Consequently, under current circumstances, replacement of these devices before the end of their useful life is unlikely to be financially attractive, so that a ban by itself on new irradiators would most likely only result in a slow decline in the stock of cesium chloride irradiators over several decades.

One strategy for making replacement of existing cesium chloride irradiators more financially attractive is to make their scrap value positive or at least nonnegative. Although the committee was not able to establish a specific scrapping cost for cesium chloride irradiators, it appears that suppliers charge approximately $35,000 to $45,000 for stand-alone disposal cost of an undamaged device (including estimates for travel, expenses, labor, shipping, and rigging charges) as indicated by MDS Nordion (2006) and another manufacturer quoted in a decommissioning plan cited by U.S. NRC staff (personal communication with M. Lowenthal,

May 16, 2007). The price might be lower to customers replacing an undamaged irradiator as part of the purchase of another device. The scrapped device might then be refurbished and sold to a user in a less developed country, the source might be reused, or the device might be stored pending availability of a disposal facility. Refurbishment and resale of a scrapped device, however, simply shifts the location of the risk, and so this approach is not favored by the committee.

The device user may also be able to dispose of it and its radiation source without a direct monetary cost through the Offsite Source Recovery Project (OSRP) administered by the National Nuclear Safety Administration. The OSRP queries source owners to assess the efforts they have made to dispose of the sources through existing disposition pathways. If owners have opportunities to dispose of the sources themselves, then the National Nuclear Safety Administration makes a decision of whether or not to recover the device. A rough estimate of the cost of recovering irradiators through the OSRP can be made based on the 2006 and 2007 irradiating device campaigns (Pearson, 2007). The cost of the contracts for these two campaigns summed to approximately $1.7 million. The campaigns recovered, or will recover, a total of 46 irradiators from 39 sites, suggesting an average of about $40,000 per irradiator. None of these costs are borne by the owners of the unwanted devices, unless they want to expedite their disposal by paying for removal and transport of the device from their premises to the OSRP facility at Los Alamos National Laboratory.

The committee was told that the OSRP is considering whether to ask the owners of unwanted sources whether they plan to simply replace the device with a new radionuclide radiation source. If a ban on new devices is not in place, simply refusing to accept old radioactive cesium chloride devices from licensees who are acquiring new radioactive cesium chloride devices avoids subsidizing the replacement (by not providing the owner with free disposal of the old device) and thus creates a disincentive to acquiring a new cesium chloride irradiator. Similarly, recovering radioactive cesium chloride devices and sources and storing them without consideration of whether alternative disposal options are available to users would encourage decommissioning of cesium chloride irradiators. The OSRP costs money, and the committee estimates that such a campaign could cost $50 million over the next decade or longer to fund the recovery of the approximately 1,300 cesium chloride devices currently in use.[4] The advantage of this approach is that the recovered irradiators and their sources would be retired not only from use in the United States but also from recycling to users in other countries that might have a less robust regulatory system.

This modification to the OSRP would have only modest effects on the retirement of cesium chloride irradiators. A more aggressive buyback policy may be desirable to speed the replacement of these irradiators. Specifically, the OSRP program could be authorized and funded to buy cesium chloride irradiators at a positive price. Either recovery or buyback would be most effective if coupled with either the ban on new cesium chloride licenses, as suggested by the committee, or an explicit requirement that participants not replace scrapped machines with new cesium chloride devices. Otherwise, the program would become a giveaway.

The most likely replacement technology for cesium chloride irradiators is the x-ray irradiator. The initial capital cost of an x-ray irradiator such as the Raycell® is approximately $180,000. In addition, the x-ray irradiators involve substantially higher maintenance costs, because the x-ray tubes must be replaced (the manufacturers told the committee that replacement should be expected about every five years) and power sources sometimes fail. A typical service contract for an x-ray irradiator such as the Raycell® currently costs about $10,600 per year. Further, relative to cesium chloride irradiators with current safety requirements, the operating costs of x-ray irradiators are likely to be somewhat higher because of the cost of electricity and the lower reliability, although the gamma irradiator has higher security, regulatory,

[4] Note that this does not include devices held by manufacturers and distributors.

and decommissioning/disposal costs. For purposes of estimating the magnitude of the buyback price necessary to induce early retirement of cesium chloride irradiators, we ignore operating costs and focus instead only on capital and maintenance costs.

The following estimates are based on the answer to the following simple question: For any given interest rate and years of remaining life in a cesium chloride irradiator, how high would the buyback price have to be to induce a switch to an x-ray irradiator? The answer to this question assumes a ban on new cesium chloride irradiators and on recharging of existing cesium chloride irradiators so that a switch would be made to an x-ray irradiator at the end of the useful life of the currently used cesium chloride irradiator.

The buyback price has two components: (1) the financial cost of purchasing the $180,000 x-ray irradiator now rather than at the end of the life of the cesium chloride irradiator, and (2) the cost of the annual service contract for each of the earlier years of use. Component 1 is simply the current purchase price of the x-ray irradiator minus the present value of the same purchase at the end of life of the cesium irradiator. Component 2 is estimated as an annual service contract payment of $10,600.

Table 10-2 shows the estimated buyback prices as a function of the real interest rate (the nominal interest rate with inflation removed) and the number of years of life remaining for the cesium chloride irradiator. The real interest rate of 6 percent is based on the 2007 prime lending rate (8.25 percent) plus 1 percent minus the 2006 inflation rate (3.25 percent). As indicated in the middle column of Table 10-2, assuming that irradiator owners face a real discount rate of 6 percent, a buyback price of approximately $90,000 would lead to the retirement of irradiators with 5 years of remaining life; a buyback price of about $158,000 would lead to the retirement of irradiators with 10 years of remaining life. Looking down the columns, one can see that inducing the retirement of devices with longer remaining useful lives would require higher buyback prices. Looking across rows, higher assumed real interest rates require higher buyback prices for short-lived devices for which the incremental cost of switching is dominated by the initial capital cost and lower buyback prices for the very long-lived devices for which switching is dominated by the longer period of higher annual maintenance costs.

If a ban on new cesium chloride irradiators were put in place, then it is likely that manufacturers would increase their investments in bringing more reliable and less costly substitutes to the market. Specifically, it would be reasonable to expect that with a larger market, the life-cycle costs of x-ray irradiators would decline. Consequently, a buyback program would become more effective over time. That is, a buyback program that currently retires devices with five years of life remaining would likely retire devices with more than five years of life remaining in the future. Further, the owner gets a newer product with greater throughput. The retiring unit has longer irradiation times because the source has diminished due to radioactive decay, whereas an x-ray unit should have a steady irradiation time throughout its lifetime. A 15-year-old cesium-137 irradiator takes 40 percent longer to reach the same dose as when it was purchased. The dollar amounts shown in Table 10-2 assume that the x-ray is the replacement technology. It is also possible that the replacement might be an irradiator (or even just a set of source pencils) that uses cesium in some other form than highly dispersible cesium chloride, assuming a more robust form were available. In this case, the incremental maintenance costs would be zero. Table 10-3 assumes that the capital cost of the replacement device is $200,000, slightly higher than the current costs of a cesium chloride irradiator. The estimates assuming a 6 percent real discount rate suggest that a buyback price of approximately $100,000 would induce the retirement of cesium chloride irradiators in between 10 and 15 years.

TABLE 10-2 Breakeven Buyback Prices Assuming X-Ray Replacement (Thousands of Dollars)

	Real Interest Rate		
Years of Life Remaining	**0.04**	**0.06**	**0.08**
5	79	90	100
10	144	158	168
15	198	208	214
20	242	245	245.
25	278	273	267

TABLE 10-3 Breakeven Buyback Prices Assuming Alternative-Form Cesium Replacement (Thousands of Dollars)

	Real Interest Rate		
Years of Life Remaining	**0.04**	**0.06**	**0.08**
5	36	51	64
10	65	88	107
15	89	117	137
20	109	138	157
25	125	153	171

Because the committee was not able to monetize the risks associated with the use of cesium chloride irradiators, or the risks associated with the use of alternative-form cesium irradiators, it does not make a specific recommendation about the magnitude of a desirable buyback price. Any buyback program must be structured carefully to ensure that no one has an incentive to buy a cesium chloride irradiator before the policy goes into place.

Cost-Benefit Perspectives

The committee's charge does not call for full cost-benefit analyses of radionuclide radiation source replacement, and it would not have been feasible for the committee to attempt such analyses. Indeed, even in attempting a cost analysis of the TSC includes several costs that are difficult to quantify, such as the psychological impact on the city and the nation, which might be greater in eventual costs than any costs from cleanup or crop loss. These costs to stakeholders range from a city's dependence on tourism being threatened for years to come, to the medical impact of chronic fear among the citizens, not to mention the costs of additional security and other proactive measures that are taken throughout society after a large disaster or attack. The committee believes that the qualitative cost-benefit framework is useful and that the committee does not have enough information to push the quantitative analysis very far. However, there is some merit in illustrating the quantitative cost-benefit approach, and so the committee provides a limited example with some notional values.

Example: There are approximately 1,300 self-contained irradiators in the United States that employ radioactive cesium chloride sources. The risk of a device being used in an RDD attack that causes severe consequences is the product of the likelihood of an attack (including accessing the source or device and weaponizing it) and the consequences of that attack, here measured just by the monetary costs for temporary loss of use and for decontamination (not the unknown psychological and long-term economic costs). The committee has no basis for establishing an absolute probability that one of these devices will be stolen and weaponized or weaponized in place, and used in an area-denial RDD attack. Government officials who carry out threat and vulnerability assessments are better positioned to evaluate those probabilities, although they are not usually assessed quantitatively. The committee can reason through a few aspects of probability on a relative basis: Better security lowers the probability of an event and therefore lowers the risk. Replacing the radionuclide radiation source with a lower hazard radiation source reduces the consequences (and may therefore lower the attractiveness of the device as a weapon). Replacing the radionuclide radiation source with a radiation generator reduces the consequence of an RDD attack using that device to zero and so reduces the risk to zero.

Consider the possibility of a high-consequence attack, hypothetically $5 billion for cleanup and other near-term economic damage, and a hypothetical cost of phasing out these devices over 20 years of $100,000 to eliminate the risk from each device through replacement. One breaks even on the expenditures versus the economic risks if the probability of a high-consequence attack is about 1 in 50 over the 20 years. That is, if one were to assess that the probability of a $5-billion–consequence RDD attack is greater than about 1 in 1,000 per year (1 in 50 is approximately equal to an annual probability of 1 in 1,010), then a $100,000 per device expenditure that eliminates the RDD risk for that device yields a net benefit. This example oversimplifies somewhat because it neglects the time value of money (i.e., does not discount future costs) and it uses hypothetical values, but it illustrates the quantitative cost-benefit approach and a possible scale to consider. Of course, if one is allocating limited resources using this approach, then a broad set of risks is assessed and resources are expended on the risk mitigation strategies with the greatest net benefit for the whole set. Assessing net benefits can be challenging and demands care in identifying and evaluating the full effects of an action, including potential international spillover from domestic actions. For example, if a U.S. ban reduces world prices for a highly hazardous radionuclide radiation source because the United States accounts for a substantial fraction of demand and world supply is price elastic, then use of the devices might increase elsewhere in the world because of a lower price. This in turn may increase the risk of an RDD attack in the United States that makes use of radiation sources from other countries. These and other considerations are part of a full cost-benefit analysis of radionuclide radiation source replacement.

Reduce Use of Other Category 1 and 2 Radionuclide Sources

Although eliminating the use of cesium chloride irradiators deserves the highest priority by policy makers, speeding the introduction of replacement technologies also deserves consideration. This, too, can be done through push, pull, and supply incentives outlined in Table 10-1.

For example, one approach for making alternative technologies more attractive is to require that licensees bear the full life-cycle costs of the radiation source. The life-cycle costs include the costs of manufacturing (borne largely outside the United States), transportation, and decommissioning (including disposal), as well as the risks associated with diversion. This approach could be put into practice through two primary mechanisms. First, the OSRP could be required to estimate the costs of safe disposal of each type of radionuclide that could be the basis for the establishment of a decommissioning fund requirement for licensees. The precedent for this approach is the decommissioning financial assurance requirements for sealed-source licensees. Second, the U.S. NRC, using a comprehensive risk approach, could set a scale of risk for different radionuclides that would be the bases for fees on either their stocks or additions to their stocks. These fees would internalize some of the risk now borne by society but not by licensees. Even if they did not fully internalize the risk, such fees would raise the cost of using radionuclides relative to alternative technologies, thereby speeding the rate of development and adoption of radiation source alternatives.

Who Bears the Burden?

As noted at the beginning of this chapter, the extent to which the user should be responsible for costs of disposal and the costs that might be incurred in association with radiological terrorism is open to debate. In the committee's view, however, current regulations do not adequately address decommissioning costs for Category 1 and 2 cesium chloride sources. One might argue that the level of the threat of radiological terrorism is beyond the control of the user and depends on government entities dealing with national security. Indeed, the threat of radiological terrorism is beyond the user's direct control. However, vulnerability is within the user's control, as is consequence, to some extent, and these are components of terrorism risk. The costs of disposal and potential terrorism are real and to decide not to internalize the costs is a decision to support use of radionuclide radiation sources rather than to discourage their use. However, users make choices based on the regulatory environment in place at the time of purchase. The licensees who already own these sources are owed a greater duty to help defray costs because society has chosen to change the regulatory environment after they made their purchase.

FINDINGS AND RECOMMENDATIONS

Finding: Nonradionuclide replacements exist for nearly all applications of Category 1 and 2 radionuclide sources (not just radioactive cesium chloride). At this time, these replacements may not all be practical or economically attractive, but most of them are improving.

Chapter 4 shows a variety accelerator systems that can be designed to operate as radiation-generator replacements for radionuclide sources. In Chapter 5, the committee explains that *self-shielded irradiators* can be operated with x-ray generators instead of radionuclides. Some x-ray-based irradiators are already commercially available, and more companies that design and manufacture x-ray generators told the committee that they are considering entering the market. As described in Chapter 6, large companies in the business of *sterilization of medical supplies and devices* operate several kinds of facilities (ethylene oxide, gamma irradiation, and electron-beam irradiation) to use the technology that is best suited to the sterilization contract. An x-ray irradiation facility can be a direct replacement for a cobalt-60 panoramic gamma irradiator, and offers both electron-beam and x-ray irradiation in one facility. The first very large scale facility for x-ray irradiation is to be built soon in Fleurus, Belgium. It is unclear whether such facilities will be cost neutral, more expensive, or less expensive per pallet of goods irradiated than similarly sized gamma irradiators. As noted in Chapter 7, linear accelerators for *radiotherapy* have almost entirely replaced cobalt-60 teletherapy devices in the United States, except for the Gamma Knife®, the use of which is still growing. The Gamma Knife® is less versatile than a linear accelerator for radiotherapy, but offers some advantages, which their competitors are trying to match with accelerators. The development of new technologies, especially in the areas of ultrasonics and x-ray sources, have provided several alternatives to *gamma radiography* in the field of nondestructive inspection. In some areas, it is likely that the use of some of the alternatives is currently limited by the availability of trained personnel and wider acceptance of the results as durable records of proper inspection, as noted in Chapter 8. Chapter 9 similarly explains that the *neutron well logging* tools that use americium-beryllium sources are beginning to see competition from accelerator fusion sources.

Finding: Neither licensees nor manufacturers now bear the full cost of liabilities related to misuse of Category 1 and 2 radiation sources, nor do they bear the costs of disposal of cesium and americium sources.

Category 1 and 2 radiation-source licensees are not required to be insured for the possible consequences of malicious use of their radiation sources. This is no different than in other sectors of our society, but it means that the costs of some liabilities are not borne by licensees. In addition, licensees of Category 1 and 2 cesium-137 and americium-241 sources in the United States do not now bear the costs of disposal of their sources because the only disposal facilities for these sources can only accept sources that come from the Department of Energy (DOE) or its predecessor, the Atomic Energy Commission. DOE has its OSRP program, which packages, transports, and stores high-risk radiation sources and devices without fee. Some licensees pay for the cost of packaging and transportation to effect the removal on their own schedule, but the cost is lower than the cost of disposal will be in an as-yet-unknown disposal facility for Greater-than-Class-C low-level waste.

Recommendation: In addition to actions related to radioactive cesium chloride, the U.S. government should adopt policies that provide incentives (market, regulatory, or certification) to facilitate the introduction of replacements and reduce the attractiveness and availability of high-risk radionuclide sources.

The committee describes several options for implementation of alternatives in this report. Among these options are to make licensees bear the full life-cycle cost of radiation sources, particularly for disposal of cesium-137 and americium-241 sources; to revise the requirements for decommissioning funds for Category 1 and 2 devices to increase the up-front costs for higher hazard sources; enhance DOE's OSRP to include a buyback of devices that still have use value, provided that the devices are replaced with lower hazard devices. The government could impose charges on all sources, or just on new sources, based on hazards or risks.

CONCLUSIONS

A variety of policies could be used to speed the replacement of Category 1 and 2 sources. Beyond a reconsideration of security requirements by the U.S. NRC, using a more comprehensive set of potential consequences, the committee views a ban on new licenses for cesium chloride irradiators as the policy most worthy of immediate consideration by policy makers. The committee also sees enabling the OSRP to recover cesium chloride irradiators more quickly as worthy of immediate consideration (as long as the old devices are not replaced with cesium chloride irradiators and the recovered devices and sources are not recycled). Buying back irradiators at a positive price to speed their replacement with alternative sources should also be considered, especially if supported by a more comprehensive assessment of risks by the U.S. NRC. Requiring that Category 1 and 2 source users establish decommissioning funds that reflect the full social costs of disposal should be considered as part of a long-term strategy for reducing the uses of radiation sources that involve net social costs.

REFERENCES

ACS (American Cancer Society). 2007. Cancer Facts & Figures 2007. Atlanta. ACS.

Adolphsen, C. 2005. Advances in normal conducting accelerator technology from the X-band linear collider program. *Proceedings of the IEEE Particle Accelerator Conference*. 2005(1590396):204-208.

American Red Cross. 2007. Isodose curve in percentages for Gammacell 1000. Unpublished data provided by the American Red Cross, 2007.

ANL (Argonne National Laboratory). August 2005. Radiological dispersal device (RDD). Human Health Fact Sheet. Accessed at http://www.ead.anl.gov/pub/doc/rdd.pdf on May 4, 2007.

ANSI/API Spec 5L. 2007. Specification for Line Pipe, 44th ed. American Petroleum Institute.

ASM International. 1988. Nondestructive Evaluation and Quality Control. American Society of Nondestructive Testing, Columbus, OH (1988), Metals Handbook Volume 17: 295-357.

Asmuss, A. 1995. Early History of X Rays. Beam Line 25 (2): 10-24.

AuBuchon J. P., L. Herschel, J. Roger, H. Taylor, P. Whitley, J. Li, R. Edrich, R.P. Goodrich. 2005. Efficacy of apheresis platelets treated with riboflavin and ultraviolet light for pathogen reduction. *Transfusion* 45:1335-1341.

Auslender, V.L.; Bryazgin, A.A.; Gorbunov, V.A.; Cheskidov, V.G.; Gornakov, I.V.; Faktorovich, B.L.; Nekhaev, V.E.; Panfilov, A.D.; Sidorov, A.V.; Tkachenko, V.O.; Tuvik, A.F.; Voronin, L.A. 2005. Industrial Electron Accelerators Type ILU. Proceedings of the Particle Accelerator Conference in Knoxville, Tennessee on 16-20 May 2005, pp. 1572-1574.

Ayan, C, Haq, S.A., Boyd, A., El-Hamawi, M., Hafez, H.H., 1999, Integration of NMR, wireline tester, core and open hole log data for dynamic reservoir properties, SPE-53273, in 11th SPE Middle East oil show and conference proceedings: Society of Petroleum Engineers.

Badruzzaman, A. 2002. Tutorial on Nuclear Logging – A Subsurface Measurement Technology, Proc. 12th Biennial American Nuclear Society RPSD Topical Meeting, April 15-18, 2002, Santa Fe, NM.

Ballas, L. K., E. Elkin, D. Schrag, and B. Minsky. 2006. Radiation therapy facilities in the United States. *International Journal of Radiation OncologyBiologyPhysics* 66:1204-1211.

Becker, A. J., J. R. Boyce, G.W. Corris, G. King III, and R. L. Bramblett, 1987. Detection of scattered x-rays from an electron linac in a borehole. *Nuclear Instruments and Methods in Physics Research* B24/25:995-998.

Beczek, K., J. W. Lewellen, A. Nassiri, and E.Tanabe. 2001. *Proceedings of the IEEE Particle Accelerator Conference* 3:2206-2208.

Benjamin, R. J., J. McCullough, P. D. Mintz, E. Snyder, W. D. Spotnitz, R. J. Rizzo, D. Wages, J. S. Lin, L. Wood, L. Corash, and M. G. Conlan. 2005. Therapeutic efficacy and safety of red blood cells treated with a chemical process (S-303) for pathogen inactivation: A Phase III clinical trial in cardiac surgery patients. *Transfusion*, Nov; 45(11):1739-49.

Berejka, A. J. 1995. Irradiation processing in the 90's: Energy savings and environmental benefits. *Radiation Physics and Chemistry* 46(4-6):429-437.

Betti, O. O., and V. E. Derechinsky, 1984. Hyperselective encephalic irradiation with linear accelerator. *Acta Neurochirurgica Suppl (Wien)* 33:385-390.

Blettner, A., D. Chauveau, and F. Gresset. 2000. Results of the first industrial applications of the new generation of imaging plates. *Proceedings of the 15th World Conference on Nondestructive Testing,* Roma, Italy, October15-21, 2000. AIPnD, Brescia, Italy. Accessed at www.ndt.net/article/wcndt00/papers/idn175/idn175.htm on November 16, 2007.

Booth, R. 1967. Rotating Neutron Target System. *IEEE Transactions on Nuclear Science* 14:938-942.

Boyce, J. R., G. King, W. Diamond, A. J. Becker, J. Doucet, and R. L. Bramblett. 1986. An electron linac as an x-ray source for measuring geological density. *Nuclear Instruments and Methods in Physics Research A* 242:507-510.

Bradley, J., D. Rees, R, Przeklasa, R. Jaitly, G. Schofield, M. Scott. 1999. *Proceedings of the IEEE Particle Accelerator Conference* 2:1010-1012.

Brandao-Mello, C.E., R. Farina, A. Rodrigues De Oliveira, M. P. Curado, J. F. Filho, and Q. C. B. Santos. 2000. Medical Follow-Up of the Radiation Accident with ^{137}Cs In Goiania—An Update (1990-1994). In Restoration of Environments Affected by Residues from Radiological Accidents: Approaches to Decision Making. Proceedings of an International Workshop held in Rio de Janeiro and Goiania, Brazil, August 29-September 2, 1994. Vienna, Austria: International Atomic Energy Agency.

Breidenback, M. Stanford Linear Accelerator Center. Personal communication with T. Raubenheimer, 2007.

Buckles, C., and P. Chipman. 1999. Ethylene Oxide User's Guide, second edition. Celanese Ltd., The Dow Chemical Company, Shell Chemical Company, Sunoco, Inc., and Equistar Chemicals, LP.

Burkhart, B. 2006. Neutron Generators and Well Logging. Presentation to the committee.

Burns, M., P. Allison, R. Carlson, J. Downing, D. Moir, and R. Shurter. 1996. Status of the Dual-Axis Radiographic Hydrodynamics Test Facility. P*roceedings of the XVII Linac Conference.*

Butch, S. H. 1996. *Blood Irradiation: A User's Guide.* Bethesda, Md.: AABB Press.

Chao, A., Tigner, M., 1999. Handbook of Accelerator Physics and Engineering. World Scientific Publishing Co.

Cassel, R., and M. N. Nguyen. 1997. *Proceedings of the Particle Accelerator Conference* 3482 Volume 3, Issue , 12-16 May 1997 Page(s):i - xviii.

CDC (Centers for Disease Control and Prevention). 2005. Division of Bacterial and Mycotic Diseases: Food Irradiation. Accessed at http://www.cdc.gov/ncidod/dbmd/diseaseinfo/foodirradiation.htm on November 16, 2007.

CDC. July 2003. Dirty Bombs. Fact Sheet. U.S. Department of Health and Human Services. Accessed at http://www.bt.cdc.gov/radiation/pdf/dirtybombs.pdf on November 16, 2007.

Chang, D., H. Vinegar, C. Morriss, and C. Straley. 1993. Effective porosity, producible fluid and permeability in carbonates from NMR logging. Paper presented at Society of Petrophysicists and Well Log Analysts 34th Annual Logging Symposium, Alberta, Canada, June 13-16, 1993 Paper A.

Chanin, D. I., and W. B. Murfin. May 1996. Contract Report. Site Restoration: Estimation of Attributable Costs from Plutonium Dispersal Accidents. Report No. SAND96-0957. Prepared by Sandia National Laboratories, Albuquerque, NM.

Chmielewski, A. G., and M. Haji-Saeid. 2004. Radiation technologies: Past, present and future. *Radiation Physics and Chemistry* 71:16–20N.

Christofilos, N. C., R. E. Hester, W. A. S. Lamb, D. D. Reagan, W. A. Sherwood, and R. E. Wright. 1964. High Current Linear Induction Accelerator for Electrons. *Review of Scientific Instruments* 35:886.

CIS-US, Inc., IBL-437 Blood Irradiator and DoseWriter. Accessed at http://www.cisusinc.com/ibl.htm on November 15, 2007.

Clarke, D. R. 1983. Ceramic materials for the immobilization of nuclear waste. *Annual Reviews of Materials Science* 13:191-218.

Clarke, D. R., P. E. D. Morgan, C. M. Jantzen,and A. B. Harke. 1981. High alumina tailored nuclear waste ceramics. *Journal of the American Ceramic Society* 64:249-258.

Cleland, M. R. 2006. Presentation to the Committee on Radiation Source Use and Replacement. Washington, DC, December 8.

Cleland, M. R.,Inventor. 1959. Voltage Multiplication Apparatus, U.S. Patent 2,875,394.

Colombo, F., A. Benedetti, F. Pozza. 1985. External stereotactic irradiation by linear accelerator. *Neurosurgery* 16:154-160.

Connell, L.W., L. C. Trost, S. W. Longley, and J. L. LaChance. May 2006. Radiological Dispersal Device (RDD) Global Protection Architecture.Prepared for the Domestic Nuclear Detection Office, Department of Homeland Security Washington, DC.

Connell, L., L. C. Trost, and S. W. Longley. May 2003. A Risk Based Approach to Radiological Dispersal Device (RDD) Analysis. Report No. SAND2003-1331, Sandia National Laboratories, Albuquerque, NM.

Cook, S. S. 1996. Selection and installation of self-contained irradiators. Blood Irradiation: A Users Guide, S. H. Butch and A. Tiehen, eds., Bethesda, MD: AABB Press.

Cook, D., A. Stassinopoulos S. Wollowitz, et al. 1998. In vivo analysis of packed red blood cells treated with S-303 to inactivate pathogens. *Blood*. 92(Suppl. 1): 503a. [Full author list not available]

Corash, L., and L. Lin. 2004. Novel processes for inactivation of leukocytes to prevent transfusion-associated graft-verus-host disease. *Bone Marrow Transplant* 33:1-7.

CRCPD. (Conference of Radiation Control Program Directors). July 11, 2007. CRCPD Assistance with Unwanted Radioactive Material. Accessed at http://www.crcpd.org/UnwantedRadMat.asp on November 15, 2007.

CRCPD. 2006. *Handbook for Responding to a Radiological Dispersal Device First Responder's Guide — The First 12 Hours.*

Creech, H.J. and A.P. O'Connell. 1981. Immunochemistry of conjugates prepared from serum albumins and acridine nitrogen mustards (ICR mutagens). *Cancer Research*. 1981;41:3844-51.

Creech, M. 2006. Considerations & Alternatives to Use of Gamma Radiation for Industrial Quality Assurance Applications. Presentation given to committee on October 26, 2006.

De Cort, M., G. Dubois, Sh. D. Fridman, M. G. Germenchuk, Yu. A. Izrael, A. Janssens, A. R. Jones, G. N. Kelly, E. V. Kvasnikova, I. I. Matveenko, I. M. Nazarov, Yu. M. Pokumeiko, V. A. Sltak, E. D. Stukin, L. Ya. Tabachny, Yu. S. Tsaturov, 1998. Atlas of Caesium Deposition on Europe after the Chernobyl Accident, EUR Report nr. 16733, Office for Official Publications of the European Communities, Luxembourg.

Deeley, C. 2001. Food irradiation: Setting new standards or a slippery slope? *Food Science and Technology* 16(2):52-55

Dever, J. P., K. F. George, W. C. Hoffman, and H. Soo. 1994. Ethylene oxide. Pp. 915-959 in *Kirk-Othmer Encyclopedia of Chemical Technology*, Vol. 9, J. I. Kroschwitz and M. Howe-Grant, eds.. New York: John Wiley & Sons

DHS (Department of Homeland Security). December 2004. National Response Plan. Accessed at http://www.dhs.gov/xlibrary/assets/NRP_FullText.pdf on March 15, 2007.

DHS. 2003. Working Group on Radiological Dispersal Device (RDD) Preparedness, Medical Preparedness and Response Sub-Group. 12/09/2003 version. Accessed at http://www1.va.gov/emshg/docs/Radiological_Medical_Countermeasures_Indexed-Final.pdf on November 16, 2007.

Diversified Technologies, Inc. 2007. DTI Power Supplies. Accessed at http://www.divtecs.com/ps_06.htm on November 16, 2007.

Dobashi, K., M. Uesaka, F. Sakamoto, T. Kaneyasu, T. Yamamoto, De Menga, J. Urakawa, T. Higo, M. Akemoto, and H. Hayano. 2005. Monochromatic tunable Compton scattering x-ray source using x-band multi-bunch linac and YAG laser circulation system. *Japanese. Journal of Applied Physics* 44:1999.

DOE (Department of Energy). 2006. Report to the U.S. Congress Under Public Law 109-58, The Energy Policy Act of 2005. Report on Alternatives to Industrial Radioactive Sources.

DOE. 1990. Interim Report of the DOE Type B Investigation Group: Cesium-137: A Systems Encapsulation to Release at Radiation Sterilizers, Inc., Decatur, Georgia. DOE/ORO--914. Oak Ridge, TN: DOE-Oak Ridge Operations Office.

DOE/U.S. NRC (Department of Energy/U.S. Nuclear Regulatory Commission). 2003. Radiological Dispersal Devices: An Initial Study to Identify Radioactive Materials of Greatest Concern and Approaches to Their Tracking, Tagging, and Disposition. Report prepared by the DOE/NRC Interagency Working Group on Radiological Dispersal Devices for the U.S. NRC and the Secretary of Energy, May 7. Accessed at http://www.nti.org/e_research/official_docs/doe/DOE052003.pdf on November 16, 2007.

Ebo, D. G., J. L. Bosmans, M. M. Couttenye, and W. J. Stevens. 2006. Haemodialysis-associated anaphylactic and anaphylactoid reactions. *Allergy*. 61:211-220.

EIA (Energy Information Administration). 2006. Electric Power Annual 2005; Revised Data. DOE/EIA-0348(2005). Accessed at http://www.eia.doe.gov/cneaf/electricity/epa/epa_sum.html on November 16, 2007.

Ellis, D. 1987. *Well Logging for Earth Scientists*. New York: Elsevier Science.

Ennever, J.F., and W. T. Speck. 1981. Photodynamic reaction of riboflavin and deoxyguanosine. *Pediatric Research* 15:956-958.

Enserink, M. 2007a. Entomology: Proven technology may get a makeover. *Science* 317(5836):312-313.

Enserink, M. 2007b. Entomology: Welcome to Ethiopia's fly factory. *Science* 317(5845):1678

Eyharts, Ph., Ph. Anthouard, J. Bardy, C. Bonnafond, Ph. Delsart, A. Devin, P. Eyl, J. Labrouche, J. Launspach, J. De Mascureau, E, Merle, A. Roques, P. Letallandier, M. Thevenot, D. Villate, and L. Voisin, 1995. Status of the Airix Induction Accelerator. *Proceedings of the IEEE Particle Accelerator Conference*, Dallas. Volume 1, Issue , 12-16 May 1997 Page(s):1254 - 1256 vol 1

Fast, L. D., G. DiLeone, J. Li, and R. Goodrich. 2006. Functional inactivation of white blood cells by Mirasol treatment. *Transfusion* 46:642-648.

Faure, J., C. Rechatin, A. Norlin, A. Lifschitz, Y. Glinec, and V. Malka. 2006. Controlled injection and acceleration of electrons in plasma wakefields by colliding laser pulses. *Nature* 444:*737.*

FDA (U.S. Food and Drug Administration). 2004. Food Allergen Labeling and Consumer Protection Act of 2004. Accessed at http://www.cfsan.fda.gov/~dms/alrgact.html on November 15, 2007.

Federal Register. January 3, 2006. Department of Homeland Security (DHS); Preparedness Directorate; Protective Action Guides for Radiological Dispersal Device (RDD) and Improvised Nuclear Device (IND) Incidents. 71(1):13414.

Ferguson, C. D., and J. O. Lubenau. 2003. Securing U.S. radioactive sources. *Issues in Science and Technology* (Fall). Accessed at http://www.issues.org/20.1/ferguson.html on November 16, 2007.

Ferguson, C. D., and W. C. Potter. 2004, The Four Faces of Nuclear Terrorism. Center for Nonproliferation Studies, Monterey Institute of International Studies, Monterey, CA. Accessed at http://cns.miis.edu/pubs/books/4faces.htm on November 16, 2007.

Ferguson, C. D., T. Kazi, and J. Perera. 2003. Commercial Radioactive Sources: Surveying the Security Risks. Occasional Paper No. 11, Center for Nonproliferation Studies, Monterey Institute of International Studies, Monterey, CA. Accessed at http://cns.miis.edu/pubs/opapers/op11/op11.pdf on November 16, 2007.

Firestone, R. B., and V. S. Shirley (eds.) 1996. *Table of Isotopes, Eighth Edition*, Vols I and II. New York: John Wiley & Sons.

Frame, P. W. 2004. A history of radiation detection instrumentation. *Health Physics* 87:111-135.

Freund, L. B. 1998. A surface chemical potential for elastic solids. *Journal of the Mechanics and Physics of Solids* 46:1835-1844.

Fry, D. W., R. B. Harvie, L. B. Mullett, and W. Walkinshaw. 1947. Travelling wave linear accelerator for electrons. *Nature* 160:351.

GAO (Government Accountability Office). 2005. NUCLEAR SECURITY: DOE Needs Better Information to Guide Its Expanded Recovery of Sealed Radiological Sources.

GAO. 2002. Diffuse Security Threats: Technologies for Mail Sanitization Exist, but Challenges Remain. Accessed at http://www.gao.gov/new.items/d02365.pdf on November 16, 2007.

Garcia Gala, J.M., A. Ramirez Payer, C. Rayon, P. Rodriguez Vicente, C. Roson, and C. Blanco. 1993. Chronic postransfusion graft-vs-host disease in a patient with non-Hodgkin's lymphoma. *Sangre.* Barcelona, Spain. 38:489-491.

Ginzton, E.L., Hansen, W.W., Kennedy, W.R. 1948. A linear accelerator, Review of Scientific Instruments. 19: 89-108.

Grady, D. E. 1982. Local Inertial Effects in Dynamic Fragmentation. Journal of Applied Physics. Vol. 53: 322.

Grass, J. A., D. J. Hei, K. Metchette, G. D. Cimino, G. P. Wiesehahn, L. Corash, and L. Lin.1998. Inactivation of leukocytes in platelet concentrates by photochemical treatment with psoralen plus UVA. *Blood* 91:2180-2188.

Gross, P. M., Dixon, L.F. 1937. Method of Sterilization. United States Patent Office. Patent 2,075,845; Patented April 6, 1937.

Haimes, Y. 2006. On the definition of vulnerabilities in measuring risks to infrastructures. *Risk Analysis* 26:293.

Haimson, J. 1975. Recent advances in high voltage electron beam injectors. *IEEE Transactions on Nuclear Science* 22:1303.

Hambleton J., T. Greenwalt, M. Viele, et al. 1999. Post transfusion recovery after multiple exposures to red blood cell concentrates (RBCS) treated with a novel pathogen inactivation (P.I.) process. *Blood* 94(Suppl.):376a (abstract). [Full author list not available]

Hardwick CC, Herivel TR, Hernandez SC, Ruane PH, Goodrich RP. Separation, identification and quantification of riboflavin and its photoproducts in blood products using high-performance liquid chromatography with fluorescence detection: a method to support pathogen reduction technology. Photochem Photobiol 2004; 80: 609–15.

Harper, F., and S. V. Musolino. 2006, Emergency response guidance for the first 48 hours after the outdoor detonation of an explosive radiological dispersal device. *Health Physics* 90(4).

Harper, F. T., S. V. Musolino, and W. B. Wente. 2007. Realistic radiological dispersal device hazard boundaries and ramifications for early consequence management decisions. *Health Physics* 93(1):1-16.

Hartmann, G. H., W. Schlegel, V. Sturm, B. Kober, O. Pastyr, and W. J. Lorenz. 1985. Cerebral radiation surgery using moving field irradiation at a linac facility. *International Journal of Radiation Oncology*Biology*Physics* 11:1185-1192.

Hayashi, H., T. Nishiuchi, H. Tamura, K. Takeda. 1993. Transfusion-associated graft-verus-host disease caused by leukocyte-filtered stored blood. *Anesthesiology*. 79:1419-21.

Health Physics Society. 2006. Continued Federal and State Action is Needed for Better Control of Radioactive Sources. Position Statement of the Health Physics Society. PS021-0, Adopted: January 2006. Accessed at http://hps.org/documents/sourcecontrol_ps021-0.pdf on July 13, 2007.

Heim, M.U., R. Munker, H. Sauer, B. Wolf-Hornung, H. Knabe, E. Holler, M. Bock, W. Mempel. 1992. Graft versus host disease with fatal outcome after administration of filtered erythrocyte concentrates. *Beitr Infusionsther*. 30:178-81.

Hendee, W. R., G. S. Ibbott, and E. G. Hendee. 2004. *Radiation Therapy Physics*, 3rd ed. Hoboken, NJ: Wiley-Liss.

Hess, N. J., F. J. Espinosa, S. D. Conradson, and W. J. Weber. 2000. Beta radiation effects in ^{137}Cs substituted pollucite. *Journal of Nuclear Materials* 281:22-33.

Hopewell Designs, Inc. 2007. Model G10 gamma beam irradiator. Accessed at http://www.hopewelldesigns.com/G10Irrad.htm on May 10, 2007.

IAEA (International Atomic Energy Agency) 2006a. Dangerous quantities of radioactive material (D-values). Emergency Preparedness and Response. IAEA, Vienna, Austria.

IAEA. 2006b. Environmental Consequences of the Chernobyl Accident and the Remediation: 20 Years of Experience. Report of the Chernobyl Forum Expert Group 'Environment'. Radiological Assessment Reports Series, Vienna, Austria.

IAEA 2005a. Categorization of Radioactive Sources, Safety Standards Series, RS-G-1.9, IAEA, Vienna, Austria.

IAEA. 2005b. Record of the CEG Workshop, Security and Safety of Radioactive Sources: Decommissioning and Replacement of Radioisotope Thermoelectric Generators (RTG), February 16-18 2005, Oslo, Norway. Accessed at http://www.iaea.org/OurWork/ST/NE/NEFW/CEG/documents/ws022005_record.pdf on November 15, 2007.

IAEA. 2005c. *Radiation Oncology Physics: A Handbook for Teachers and Students*, E. B. Podgorsak, ed. Vienna, Austria: IAEA.

IAEA. 2004a. Code of Conduct on the Safety and Security of Radioactive Sources. IAEA, Vienna, Austria.

IAEA. 2004b. Gamma Irradiators for Radiation Processing. Brochure. IAEA, Vienna, Austria.

IAEA. 2003a. International Conference on Security of Radioactive Sources. Findings of the President of the Conference. International Conference on Security of Radioactive Sources, March 10-13, 2003, Vienna, Austria. Accessed at http://www-ns.iaea.org/downloads/rw/meetings/rdd_findings.pdf on November 15, 2007.

IAEA. 2003b. Security of Radioactive Sources: Interim Guidance for Comment. IAEA Tecdoc 1355. IAEA, Vienna, Austria.

IAEA. September 2002. Inadequate Control of World's Radioactive Sources. IAEA Press Release. Accessed at http://www.iaea.org/NewsCenter/PressReleases/2002/prn0209.shtml on November 16, 2007.

IAEA. 2001. Measures to Improve the Security of Nuclear Materials and Other Radioactive Materials. Board of Governor's General Conference, August,14, 2001in Stockholm, Sweden.

IAEA. 1991. The International Chornobyl Project: Technical Report. IAEA, Vienna, Austria.

IAEA. September 1988. The Radiological Accident in Goiânia. STI/PUB/815. IAEA, Vienna, Austria.

Inspecta. 2007. Radiography with beatron up to 250 mm of steel. Accessed at http://www.inspecta.se/downloads/produktblad/010_on November 15, 2007.

International System of Units (SI). 8th ed. Paris, France: Organisation Intergouvernementale de la Convention du Mètre; 2006

IPCS (International Programme on Chemical Safety). 2001. Ethylene oxide. International Chemical Safety Cards. Accessed at http://www.cdc.gov/niosh/ipcsneng/neng0155.html. on November, 16, 2007.

Jacobsen, S., D. May, J. Grant, and J. Little. 2006. Producibility prediction in gas sands through effective integration of NMR, resistivity and porosity log data. Paper I, presented at Society of Petrophysicists and Well Log Analysts 47th Annual Logging Symposium, June 4-7.

Jammal, R. 2007. Verbal Communication with the Committee on Radiation Source Use and Replacement, January 9.

Janatpour, K., L. Denning, K. Nelson, B. Betlach, M. MacKenzie, and P. Holland. 2005. Comparison of x-ray vs. gamma irradiation of CPDA-1 red cells. *Vox Sanguinis* 89:215-219.

JME Ltd. 2007. Portable X-ray Betatron. Accessed at http://www.jme.co.uk/files/JME%20Datasheet%20PXB6.pdf on November 15, 2007.

Johns, H. E., and J. R. Cunningham. 1984. *The Physics of Radiology*. Springfield, IL: Thomas.

Johns, H. E., L. M. Bates, and E. R. Epp. 1951. 1000-Curie cobalt-60 units for radiation therapy, *Nature* 168:1035-1036.

Jolie, J., W. Mondelaers, P. Cauwels, B. Masschaele, M. Dierick, P. Lahorte, S. Baechler, and T. Materna. 1998. On the construction of a tunable high-energy photon source. *Radiation Physics and Chemistry* 51:413-421.

Jongen, Y., M. Abs, F. Genin, A. Nguyen, and J. M. Capdevila. 1993. The Rhodotron, a new 10 MeV, 100 kW, cw metric wave electron accelerator. *Nuclear Instruments and Methods in Physics Research, Section B: Beam Interactions with Materials and Atoms* 79:865-870.

Journal of American Dental Association. December 1991. 122:

Kapetanakos, C., K. L. Len, T. Smith, S. J. Marsh, P. Loschialpo, D. Dialetis, and J. Mathew. 1993. The Naval Research Laboratory Modified Betatron Accelerator and assessment of its results. *Physics of Fluids* 5:2295-2313

Kaplan, S., and J. B. Garrick. 1981. On the quantitative definition of risk. *Risk Analysis* 1(1):11-27.

Kaplin, V.V., S. R. Uglov, O. F. Bulaev, V. J. Goncharov, A. A. Voronin, M. A. Piestrup, and C. K. Gary. 2002. Thin betatron radiators for more efficient x-ray generation. *Review of Scientific Instruments* 73:63.

Kelly, H. 2002. Testimony of Dr. Henry Kelly, president of the Federation of American Scientists, before the Senate Committee on Foreign Relations, March 6.

Kerst, D. W. 1940. Acceleration of electrons by magnetic induction. *Physical Review* 58:841.

Khan, F. 2003. *The Physics of Radiation Therapy*, 3rd ed. Baltimore, MD: Lippincott, Williams and Wilkins.

King, G., A. L Becker, G. W. Cords, J.R. Boyce, and R. L. Bramblett. 1987. Density Logging using an Electron Accelerator as the X-ray Source. *Nuclear Instruments and Methods in Physics Research B* 24/25:990-994.

Kirk, R. 2006. X-ray as a substitute for Class 2 self-contained gamma irradiators. Presentation to the Committee on Radiation Source Use and Replacement. Washington, DC, December 8.

Kjellberg, R. N., A. Shintani, and A. G. Frantz. 1968. Proton beam in acromegaly. *New England Journal of Medicine* 278:689-695.

Knoll, G. F. 2000. *Radiation Detection and Measurement.* New York: John Wiley & Sons.

Komareni, S. and R. Roy, 1983. Hydrothermal Reaction and Dissolution Studies of CsAlSi5O12 in Water and Brines. Journal of the American Ceramic Society, vol 66: 471-474.

Kunstad, P. 2001. Economic and Technical Considerations in Food Irradiation. pp 415-442. In: R. A. Molins, (ed.) Food Irradiation-Priciples and Applications. Wiley-Interscience, New York.

LaMontagne, A. D., and K. T. Kelsey. 1998. Ethylene oxide. Pp. 1145-1155 in *Environmental and Occupational Medicine*, 3rd ed., W. N. Rom, (ed.). Philiadelphia: Lippincott-Raven.

Larsson, B. K., L. Leksell, B. Rexed, P. Sourander, W. Mair, and B. Andersson. 1958. The high energy proton beam as a neurosurgical tool, *Nature* 182:1222-1223.

Larsson, B. K., K. Liden, and B. Sorby. 1974. Irradiation of small structures through intact skull. Acta Radiological: Therapy, Physics, Biology 13:513-534.

Lauffer, L., K.H. Weber, F. Hucho. 1979. Acetylcholine receptor. Binding properties and ion permeability response after covalent attachment of the local anaesthetic quinacrine. *Biochimica et Biophysica Acta*. 587:42-8.

Lawrence, J. H., C. A. Tobias, J. L. Born, C. C. Wang, and J. H. Linfoot. 1962. Heavy particle irradiation in neo-plastic and neurologic disease. *Journal of Neurosurgery* 19:717-722.

Leksell, L. 1951. The stereotaxis method and radiosurgery of the brain. *Acta Chirurgica Scandinavica* 102:316-319.

Leksell, L. 1968. Cerebral radiosurgery I. Gamma thalomotomy in two cases of intractable pain. *Acta Chirurgica Scandinavica* 134:585-595.

Lenci, S., A. Balkcum, H. Bohlen, A. Mizuhara, E. Wright, Y. Li, and R. Tornoe. 2004. RF sources for 3rd & 4th generation light sources. Pp. 566-569 in *Proceedings of the 2004 FEL Conference, Trieste, Italy. August 29-September 3. Accessed at http://www.elettra.trieste.it/fel2004/proceedings/ on April 3, 2008. .*

Lewis, R. J. (2003). *Ethylene Oxide*. New York: John Wiley & Sons.

Luban, N. L., D. Drothler, G. Moroff, and R. Quinones. 2000. Irradiation of platelet components: inhibition of lymphocyte proliferation assessed by limiting-dilution analysis. *Transfusion* 40:348-352.

Lubenau, J. O. 1999. A century's challenges: Historical overview of radiation sources in the United States. *IAEA Bulletin* 41(3):2.

Lubenau, J. O., and D. G. Strom. 2002. Safety and security of radiation sources in the aftermath of 11 September 2001. *Health Physics* 83:155-164.

Lutz, W., K. R. Winston, and N. Maleki. 1988. A system for stereotactic radiosurgery with a linear accelerator. *International Journal of Radiation Oncology*Biology*Physics* 14:373-381.

Maciszewski, W., and W. Scharf. 2004. Particle accelerators for radiotherapy: Present status and future. *Physica Medica* 20:137-146.

MacLaren, I., J. Cirre, and C. B. Ponton. 1999. Hydrothermal synthesis of pollucite ($CsAlSi_2O_6$) powders. *Journal of the American Ceramics Society* 82:3242-3244.

Malvadkar, S. B., and E. L. Parsons. 2002. Analysis of Potential Power Sources for Inspection Robots in Natural Gas Transmission Pipelines. Topical Report DE-FC26-01NT41155. National Energy Technology Laboratory. Accessed at http://www.netl.doe.gov/technologies/oil-gas/publications/td/parsons%20malvadkar%20report.pdf on May 7, 2007.

Marshall, E. 1984. Juarez: An unexpected radiation accident. *Science* 223:1152-1154.

Masefield, J. 2004. Reflections on the evolution and current status of the radiation industry. *Radiation Physics and Chemistry* 71, Issues 1-2 (September-October): 9-16

McCollough, J., D. H. Vesole, R. J. Benjamin, S. J. Slichter, A. Pineda, E. Snyder, E. A. Stadtmauer, I. Lopez-Plaza, S. Coutre, R. G. Strauss, L. T. Goodnough, J. L. Fridey, T. Raife, R. Cable, S. Murphy, F. Howard, K. Davis, J. Lin, P. Metzel, L. Corash, A. Koutsoukos, L. Lin, D. H. Buchholz, M. G. Conlan. 2004. Therapeutic efficacy and safety of platelets treated with a photochemical process for pathogen inactivation: The SPRINT trial. *Blood* 104:1534-1541.

McCullough, J., H. A. Perkins, and J. Hansen. 2006. The National Marrow Program with emphasis on the early years. *Transfusion* 46:1248-1255.

MDS Nordion. 2006. Gammacell® 1000 Elite/3000 Elan. Brochure. Accessed at http://www.mds.nordion.com/documents/elibrary/blood-research-irradiators/Gammacell/Gammacell%201000-3000%20Sales%20Brochure.pdf on November 15, 2007.

MDS Nordion. 2002. JS-10000 Hanging Tote Irradiator and Pallet Irradiators. Brochures. Accessed at http://www.theratronics.com/documents/elibrary/sterilization/pallet-irradiators/Pallet_Brochure.pdf on November 15, 2007.

Medalia, J. 2004. Terrorist "Dirty Bombs": A Brief Primer. CRS Report for Congress. Order Code RS21528, Congressional Research Service, Foreign Affairs, Defense, and Trade Division, updated April 1, 2004. Accessed at http://www.fas.org/spp/starwars/crs/RS21528.pdf on November 16, 2007.

Meissner, J., M. Abs, M. R. Cleland, A. S. Herer, Y. Jongen, F. Kuntz, and A. Strasser. 2000. X-ray treatment at 5 MeV and above. *Radiation Physics and Chemistry* 57:647-651.

Miller, C. W. 1953. Traveling wave linear accelerator for x-ray therapy. *Nature* 171:297-299.

Miller, R. B., G. Loda, R. C. Miller, R. Smith, D. Shimer, C. Seidt, M. MacArt, H. Mohr, G. Robison, P. Creely, J. Bautista, T. Oliva, L. M. Young, and D. DuBois. 2003. A high-power electron linear accelerator for food irradiation applications. *Nuclear Instruments and Methods on Physics Research, B: Beam Interactions with Materials and Atoms* 211:562-570.

Mimura, H., K. Iijima., and K. Akiba, K. 1997. Leaching behavior and surface alteration of cesium aluminum silicate under static and dynamic conditions. *Journal of Nuclear Science and Technology* 34:269-276.

Mondelaers, W., P. Lahorte, B. Masschaele, and P. Cauwels. 2000. Transient beam loading compensation in low-energy pulsed linear electron accelerators. Presented at the 20th International Linac Conference, Monterey, California, August 21-25.

Moore, G. M. 2003. Radiological Weapons: How Great Is the Danger? UCRL-ID-154879. U.S. Department of Energy, Lawrence Livermore National Laboratory.

Moroff, G., and N. L. Luban. 1997. The irradiation of blood and blood components to prevent graft-versus-host disease: technical issues and guidelines. *Transfusion Medicine Reviews* 11:15-26.

Morrison, R. 1989. An Economic Analysis of Electron Accelerators and Cobalt-60 for Irradiating Food. Technical Bulletin No. 1762. Commodity Economics Division, Economic Research Service, U.S. Department of Agriculture, Washington, DC.

Musolino, S. V., and F. T. Harper. 2006. Emergency response guidance for the first 48 hours after the outdoor detonation of an explosive radiological device. *Health Physics* 90:377-385.

NASA (National Aeronautics and Space Administration). March 2006. Mars Scout 2011 AO Radioisotope Heater Unit (RHU) Information Summary. Accessed at http://mars-scout.larc.nasa.gov/PDF_FILES/Scouts2011AOannexRHUs4-19.pdf on November 15, 2007.

NASA. 2004. April 2004. Discovery AO Radioisotope heater Unit (RHU) Information Summary. Accessed at http://discovery.larc.nasa.gov/PDF_FILES/LARHU_Descrip_final.pdf on November 15, 2007.

National Research Council. 2007. *U.S.-Russian Collaboration in Combating Radiological Terrorism*. Washington, DC: National Academies Press.

National Research Council. 2006a. *Improving the Regulation and Management of Low-Activity Radioactive Wastes*. Washington, DC: National Academies Press.

National Research Council. 2006b. *Going the Distance? The Safe Transport of Spent Nuclear Fuel and High-Level Radioactive Waste in the United States*. Washington, DC: National Academies Press.

National Research Council. 2006c. *Health Risks from Exposure to Low Level Ionizing Radiation: BEIR VII, Phase 2*. Washington, DC: National Academies Press.

National Research Council. 2003. *Improving the Scientific Basis for Managing DOE's Excess Nuclear Materials and Spent Nuclear Fuel*. Washington, DC: National Academies Press.

National Research Council. 2002. *Making the Nation Safer: The Role in Science and Technology in Countering Terrorism*. Washington, DC: National Academies Press.

National Research Council. 1983. *Risk Assessment in the Federal Government: Managing the Process*. Washington, DC: National Academies Press.

NCHS (National Center for Health Statistics). 2006. *Health, United States, 2006*, with *Chartbook on Trends in the Health of Americans*. Hyattsville, MD: NCHS.

NCRP (National Council on Radiation Protection and Measurements). 2006. Cesium-137 in the Environment: Radioecology and Approaches to Assessment and Management. Bethesda, MD: NCRP

NCRP. 2005. Radiation Protection for Particle Accelerator Facilities. NCRP Report No. 144. Issued December 31, 2003, Revised March 4, 2005. Bethesda, MD: NCRP.

NCRP. 2001. Management of Terrorist Events Involving Radioactive Material. Report No. 138. Bethesda, MD: NCRP.

NCRP. 1984. Neutron Contamination from Medical Electron Accelerators. Report No. 79. Bethesda, MD: NCRP.

NCRP. 1977. Radiation Protection Guidelines for 0.1-100 MeV Particle Accelerator Facilities. Report No. 51. Bethesda, MD: NCRP.

NIOSH (National Institute for Occupational Safety and Health). 2004. Worker Notification Program—Sterilization of Medical Instruments and Treatment of Spices (Ethylene Oxide). Accessed at http://www.cdc.gov/niosh/pgms/worknotify/EthyleneOxide.html on November 15, 2007.

NIOSH. 2000. Preventing Worker Injuries and Deaths from Explosions in Industrial Ethylene Oxide Sterilization Facilities. Cincinnati, OH: NIOSH, U.S. Environmental Protection Agency, and the Ethylene Oxide Sterilization Association. Accessed at http://www.cdc.gov/niosh/2000119.html on November 16, 2007.

Norton, G. A., and G. M. Klody. 1997. P. 109 in *Proceedings of the Fourteenth International Conference on the Applications of Accelerators in Research and Industry*, J. L. Duggan and I. L. Morgan, eds. New York: AIP Press.

NRPB (National Radiation Protection Board of the United Kingdom). 1996. Review of Decontamination and Clean-up and Techniques for Use in the UK Following Accidental Releases of Radioactivity to the Environment. DOE/RAS/96.009 (NRPB, R-288).

Nucleonics Week. 1999. France wants ban on radioactive smoke detectors after truck fire. October 28.

ORTEC. 2007. Choice of Detector Shielding, Figure 32. Accessed at http://www.ortec-online.com/detectors/photon/b6.htm on November 16, 2007.

Paté-Cornell, M. E., and S. D. Guikema. 2002. Probabilistic modeling of terrorist threats: A systems analysis approach to setting priorities among countermeasures. *Military Operations Research* 7(4):5-23.

Pearson, M. W. 2007. Off-Site Source Recovery Project. Presentation to National Research Council Committee on Radiation Source Use and Replacement, January 24, 2007, Irvine, CA.

Pelszynski, M.M., G. Moroff N. L. Luban, B. J. Taylor, and R. R. Quinones. 1994. Effect of gamma irradiation of red blood cell units on T-cell inactivation as assessed by limiting dilution analysis: Implications for preventing transfusion-associated graft-versus-host disease. *Blood* 83:1683-1689.

Plappert, U., K. Raddatz, S. Roth, and T. M. Fliendner. 1995. DNA damage detection in man after radiation exposure—the comet assay—its possible application for human biomonitoring. *Stem Cells* 13:215–222.

Podgorsak, E. B., A. Olivier, M. Pla, J. Hazel, J., A. de Lotbinière, and B. Pike. 1987. Physical aspects of dynamic stereotactic radiosurgery. *Applied Neurophysiology* 50:263-268.

Pottier, J. 1989. A new type of rf electron accelerator: The rhodotron. *Nuclear Instruments and Methods in Physics Research Section B: Beam Interactions with Materials and Atoms* 40/41:943-945.

Reichmuth, B., S. Short, and T Wood. 2005. Economic Consequences of a Rad/Nuc Attack: Cleanup Standards Significantly Affect Cost.PNNL-SA-45256. Richland, WA: Pacific Northwest National Laboratories.

REVISS. 2003. A Cs-137 Source Pricing Schedule from REVISS for 2003.

Reyes-Mena, A., C. Jensen, E. Bard, D. Clark Turner, K. G. Erdmann, Q. Qiu, B. Gao, J. Lu, O. Zhou. 2005. Miniature x-ray tubes utilizing carbon-nanotube-based cold cathodes. International Centre for Diffraction Data. *Advances in X-ray Analysis* 48. 204-209. Accessed at http://www.moxtek.com/PDF/Publications/MINIATURE_X-RAY_TUBES_UTILIZING.pdf on November 16, 2007.

Rising, D. 2006. Al Qaeda in Iraq beckons nuclear scientists. Associated Press. September 29, 2006. Accessed at http://www.boston.com/news/world/articles/2006/09/29/al_qaeda_in_iraq_beckons_nuclear_scientists/ on November 20, 2007.

Roessler, G. 2007. Why ^{210}Po? *Health Physics News*. XXV, 2, February.

Roscoe, B.A., J.A. Grau, R.A. Manente, C.L. Melcher, C.A. Peterson, J.S. Schweitzer, and C. Stoller. 1991. Use of GSO for inelastic gamma-ray spectroscopy measurements in the borehole. IEEE Nuclear Science Symposium and Medical Imaging Conference, 2-9 Nov 1991: 1098-1101.

Rosoff, H., and D. von Winterfeldt. 2007. A risk and economic analysis of dirty bomb attacks on the ports of Los Angeles and Long Beach. *Risk Analysis* 27(3):533-46.

Rutala, W. A., and D. J. Weber. 2001. New disinfection and sterilization methods. *Emerging Infectious Diseases* 7(2):348-353.

Sampayan, S., G. Caporaso, Y.-J. Chen, S. Hawkins, C. Holmes, J. McCarrick, S. Nelson, W. Nunnally, B. Poole, M. Rhodes, D. Sanders, J. Sullivan, L. Wang, J. Watson. 2005. Development of a compact radiography accelerator using dielectric wall accelerator technology. *Particle Accelerator Conference* 16(20):716-718.

Setser, J. L. 1990. Leakage of an Irradiator Source—The June 1988 Georgia RSI Incident. NUREG-1392. Rockville, MD: U.S. NRC.

Slack, J., G. R. Malkoske, and L. Norton. 2003. Cobalt-60 production in CANDU® power reactors. Paper presented at the Symposium on Nuclear Energy — SIEN 2003, October 22-25, 2003, Bucharest, Romania.

Smith, M. 2006. Differences in Assumptions Imbedded in Presentations by Sterigenics International and Ion Beam Applications (IBA) to the National Academy of Sciences Committee on Radiation Source Use and Replacement.

Snyder, E., T. Raife, L. Lin, G. Cimino, P. Metzel, M. Rheinschmidt, L. Baril, K. Davis, D. H. Buchholz, L. Corash, and M. G. Conlan. 2004. Recovery and life span of ^{111}indium-radiolabeled platelets treated with pathogen inactivation with amotosalen HCl (S-59) and ultraviolet a light. *Transfusion* 44:1732-1740.

Sterigenics, Inc. 2007. Sterilization Alternatives: Ethylene Oxide. Accessed at http://www.sterigenics.com/library on May 14, 2007.

Sullivan, M. T., R. Cotten, E. J. Read, and E. L. Wallace. 2007. Blood collection and transfusion in the United States in 2001. *Transfusion* 47:366-368.

Tanabe, E., M. Borland, M. C. Green, R. H. Miller, L. V. Nelson, J. N. Weaver, and H. Wiedemann. 1989. *Proceedings of the 1989 Linear Accelerator Conference*. pp 106.

TCPA (Texas Comptroller of Public Accounts). July 1998. El Cobalto. In *Bordering the Future: Challenge and Opportunity in the Texas Border Region*. Accessed at http://www.window.state.tx.us/border/ch09/cobalto.html on January, 10, 2008.

U.S. Cancer Statistics Working Group. 2006. United States Cancer Statistics: 2003 Incidence and Mortality. Atlanta: U.S. Department of Health and Human Services, Centers for Disease Control and Prevention and National Cancer Institute.

U.S. Census Bureau. U.S. POPClock Projection. Accessed at http://www.census.gov/population/www/popclockus.html on November 12, 2007.

U.S. Chemical Safety and Hazard Investigation Board. 2006. Investigation Report: Sterigenics (4 Employees Injured). Accessed at http://www.csb.gov/completed_investigations/docs/CSBSterigenicsFinalReport.pdf on November 17, 2007.

U.S. Ecology Washington, INC. 2007. Schedule of Charges Effective May 1, 2007 Schedule A, Fifth Revision.

U.S. NRC (United States Nuclear Regulatory Commission). 2007a. Fiscal Year 2006 Interim Inventory of Radioactive Sources Data Analysis. Rockville, MD: U.S. NRC.

U.S. NRC. 2007b. NRC Regulatory Issue Summary 2007-23 Date for Operation of National Source Tracking System, September 11, 2007. Rockville, MD: U.S. NRC.

U.S. NRC. 2007c. Appendix B: Consideration of Terrorist Attacks on the Proposed Pa'ina Irradiator. Amended to Draft Environmental Assessment Related to the Proposed Pa'ina Hawaii, LLC Underwater Irradiator in Honolulu, Hawaii, June 1, 2007.

U.S. NRC. 2006a. Radiation Source Protection and Security Task Force Report. August 13. Rockville, MD: U.S. NRC.

U.S. NRC. 2006b. Fiscal Year 2005 Interim Inventory of Radioactive Sources Data Analysis. March. Rockville, MD: U.S. NRC.

U.S. NRC. 2005a. Occupational Radiation Exposure at NRC Licensed Facilities. Thirty-Eighth Annual Report. Rockville, MD: U.S. NRC.

U.S. NRC. 2005b. Order Imposing Increased Controls. EA 05-090. November. Rockville, MD: U.S. NRC.

U.S. NRC. 2004. Strategic Plan: Fiscal Years 2004-2009. NUREG-1614, Vol. 3. Rockville, MD: U.S. NRC.

U.S. NRC. 2003. Dirty bombs. Fact Sheet. Office of Public Affairs, Rockville, MD. Accessed at http://www.nrc.gov/reading-rm/doc-collections/fact-sheets/dirty-bombs.pdf on November 16, 2007.

U.S. NRC. 2001. Atomic Safety and Licensing Board Panel. Program Management Information. Gray Star Inc. Denial of Application for Sealed Source Registration Certificate (STP-01-018). Agency Document Access and Management System, Accession No. ML010710172, March 9, 2001.

U.S. NRC. 1995. Branch Technical Position on Concentration Averaging and Encapsulation. January 17. Rockville, MD: U.S. NRC.

U.S. NRC. 1989. Recent Safety-Related Incidents at Large Irradiators. Information Notice No. 89-82. Rockville, MD: U.S. NRC.

U.S. NRC. 1982. Working Safely in Gamma Radiography. NUREG/BR0024. Washington D.C. : U.S. NRC.

Valant-Spaight, B., W. Han, W. Guo, and W. Schultz. 2006. Field examples with a slim LWD density/neutron instrument containing a californium-252 neutron source and three neutron detectors. Paper presented at Society of Petrophysicists and Well Log Analysts 47th Annual Logging Symposium, June 4-7, Veracruz, Mexico.

van Ankeren, S.C., D. Murray, and R. E. Meyn. 1988. Induction and rejoining of gamma-ray-induced DNA single- and double-strand breaks in Chinese hamster AA8 cells and in two radiosensitive clones. *Radiation Research* 116:511-525.

van Laar, J.M., and A. Tyndal. 2006. Adult stem cells in the treatment of autoimmune diseases. *Rhematology (Oxford).* epub ahead of print, PMID: 16777856.

van Popta, J., P. Hofstra, and S. van Houwelingen. 2004. An Advanced Evaluation Method for Laminated Shaly Sands Including Uncertainty and Sensitivity. Society of Petrophysicists and Well Log Analysts 45th Annual Logging Symposium, June 6-9, 2004.

van Tuyle, G. J., T. L. Strub, H. A. O'Brien, C. F. V. Mason, and S. Gitomer. September 2003. Reducing RDD Concerns Related to large Radiological Source Applications. Report No. LA-UR-03-6664, Los Alamos National Laboratories, Los Alamos, NM.

Veksler, V. 1944. A New Method of the Acceleration of Relativistic Particles. *Proceedings Academy of Sciences of the USSR* 43:346

Vlieks, A. E., W. R. Fowkes, R. J. Loewen, and S. G. Tantawi. 1998. High power rf component testing for the NLC. Presented at 19th International Linear Accelerator Conference (LINAC 98), August 23-28, 1998, Chicago, IL. SLAC-PUB-7938.

Vlieks, A.E., Akre, R., Caryotakis, G., DeStefano, C., Frederick, W.J., Heritage, J.P., Luhmann, N.C., Jr., Martin, D., and Pellegrini, C., "Recent Measurements and Plans for the SLAC Compton X-Ray Source," Proceedings of the 7th Workshop on High Energy Density and High Power RF, AIP, 807, 481-490, June 13-17, 2005, Kalamata, Greece.

Westenskow, G. A., and J. M. J. Madey. 1984. Microwave electron gun. *Laser and Particle Beams* 2(Pt 2):223.

White, A., D. Burns, and T. W. Christensen. 2006. Effective terminal sterilization using supercritical carbon dioxide. *Journal of Biotechnology* 123:504-515

WHO (World Health Organization). 2005. Chornobyl: The true scale of the accident. Joint News Release. WHO/IAEA/UNDP.Accessed at http://www.who.int/mediacentre/news/releases/2005/pr38/en/index.html on September 2005.

Williams, J.R., Thwaites, D.I., "Radiotherapy physics in practice", Oxford University Press, Oxford, UK (2000).

Willis, H. H., A. R. Morral, T. K. Kelly, and J. J. Medby. 2005. Estimating Terrorism Risk. RAND Center for Terrorism Risk Management Policy, Santa Monica, CA.

Xintek Inc. 2004-2006.Current X-ray Technology. Accessed at http://www.xintek.com/products/xray/index.htm on November 15, 2007.

Yaris, L. 2003. Computer identify T cell turn-ons. *Berkeley Lab View*. Accessed at http://www.lbl.gov/Publications/Currents/Archive/Oct-03-2003.html on January 30, 2008.

Yamamoto, T., K. Dobashi, T. Natsui, M. Uesaka, E. Tanabe, T. Higo, S. Fukuda, M. Akemoto, M. Yoshida. 2006. Design of 9.4 GHz 950 keV X-band linac for nondestructive testing. Prepared for European Particle Accelerator Conference (EPAC 06), Edinburgh, Scotland, 26-30 Jun 2006. Published in *Edinburgh 2006, EPAC* 2358-2360.

Zimmerman, P. D., and C. Loeb. 2004. Dirty bombs: The threat revisited. *Defense Horizons* 38:1−11. Accessed at http://hps.org/documents/RDD_report.pdf on November 15, 2007..

Zull, L. 1996. Trip Report – Safety of Cesium and Strontium Capsules at Hanford. Defense Nuclear Facilities Safety Board, June 7.

APPENDIX A

BIOGRAPHICAL SKETCHES OF COMMITTEE MEMBERS

Theodore L. Phillips, Chair, is the Wun-Kon Fu Distinguished Professor of Radiation Oncology at the University of California at San Francisco (UCSF). Dr. Phillips was previously associate director of the UCSF Cancer Center and chairman of the department of radiation oncology. He is a past president of both the Radiation Research Society and the American Society for Therapeutic Radiology and Oncology. Dr. Phillips received the O. H. Perry Pepper Research Prize, the Gold Medal of the Gilbert H. Fletcher Society, the Janeway Medal of the American Radium Society, the Del Regato Award from Indiana University, the C. Chester Stock Award from Memorial Sloan Kettering Cancer Center, the Gold Medal of the American Society for Therapeutic Radiology and Oncology, and the Gold Medal of the American College of Radiology. He is a member of the Institute of Medicine. He has served on several committees for the National Research Council and was a member of the Board on Radiation Effects Research. He sits on the board of directors for Intraop Medical, which makes a device for intraoperative radiation therapy. Dr. Phillips received his Sc.B. in chemistry from Dickinson College and his M.D. from the University of Pennsylvania School of Medicine.

Everett E. Bloom is associate director of the Center for Materials Processing at the University of Tennessee, Knoxville. Prior to joining the university, he had a 42-year career at Oak Ridge National Laboratory culminating as director of the laboratory's Metals and Ceramics Division. He is an expert in materials science, particularly in nuclear applications. Dr. Bloom received the Department of Energy's Distinguished Associate Award for his role in the U.S. DOE Fusion Materials Program. He has authored more than 200 papers and holds several patents. He serves on the South Dakota School of Mines and Technology Academic Advisory Board and received its Distinguished Alumni Award in 2003. He is currently on the National Research Council's National Materials Advisory Board. Dr. Bloom earned his B.S. in metallurgical engineering from the South Dakota School of Mines and Technology and his M.S. and Ph.D. in metallurgical engineering from the University of Tennessee, Knoxville.

David R. Clarke is a professor in the Materials Department of the University of California at Santa Barbara, where he has worked since 1990. He previously worked at the IBM Research Division, the Massachusetts Institute of Technology, Rockwell International Science Center, the University of California at Berkeley, and the National Physical Laboratory in England. Dr. Clarke's expertise is in the relationship between material microstructure and material properties, materials design and processing, and nondestructive evaluation. Much of his research has focused on electronic and structural ceramics, but he has also worked on extending the life of gas-turbine engines and developing waste forms for radioactive waste disposal. Dr. Clarke has received numerous awards, including the Sosman Memorial Award from the American Ceramics Society and the Alexander von Humboldt Foundation Senior Scientist Award. He is an academician of the International Academy of Ceramics, a fellow of the American Physical Society and the American Ceramics Society, and a member of the National Academy of Engineering. Dr. Clarke is author or coauthor of more than 350 papers and holds 6 patents. He has served as member or chair on four previous committees of the National Research Council.

Dr. Clarke received his B.Sc. in applied sciences from the University of Sussex, England, and his Ph.D. in physics from Cavendish Laboratory at the University of Cambridge.

Leonard W. Connell is a distinguished member of the technical staff in the National Security Studies Department of the Systems Analysis Group at Sandia National Laboratories in Albuquerque, New Mexico, where he has worked for 26 years. He is a technical advisor to senior laboratory management on nuclear weapons, nuclear terrorism, and unconventional nuclear warfare. Prior to joining Sandia, he was an officer in the U.S. Navy, teaching at the Naval Nuclear Power School. Dr. Connell's work has ranged from aerodynamics of reentry vehicles and space nuclear power and propulsion for Mars exploration to pulsed laser effects and countermeasures for dealing with radiological terrorism. He has led studies at Sandia evaluating the security of gamma irradiators and the risks from radiological attacks using high-intensity radiation sources. Dr. Connell received his B.S. in mechanical engineering from Michigan Technological University and his Ph.D. in nuclear engineering from the University of New Mexico.

Robin P. Gardner is director of the Center for Engineering Applications of Radioisotopes and professor of nuclear and chemical engineering at North Carolina State University. He is an expert in industrial and medical radiation and radioisotope measurement applications. Using experiments and simulations, he works to optimize design of applications using radioisotope tracers, nuclear gauges, nuclear analyzers, nuclear oil-well logging devices and computed tomography devices. Dr. Gardner received his B.Ch.E. and M.S. in chemical engineering from North Carolina State University and his Ph.D. in fuel technology from Pennsylvania State University.

C. Richard Liu is a professor of electrical engineering at the University of Houston. He joined the department in 1988 as a postdoctoral research associate. Dr. Liu's research is in electrical properties of rocks, well logging, microwave passive and active networks, radiofrequency circuit design, radio systems and networking, numerical computation of electromagnetic scattering and propagation, hardware and software design of ground-penetrating radar systems, electromagnetic tomography, and microwave telecommunication systems. He is a member of the Research Management Committee of the Texas Department of Transportation. Dr. Liu is also an active member of the Institute of Electrical and Electronic Engineers, the Environmental and Engineering Geophysics Society, and the Society of Professional Well Logging Analysts. Dr. Liu received his B.S., M.S., and Ph. D. degrees in electrical engineering from Jiaotong University, Xian, China, in 1982, 1984, and 1988, respectively.

Ruth McBurney is the executive director of the Conference of Radiation Control Program Directors. She has worked in the field of radiation protection for over 25 years, having served previously as director of the Division of Licensing, Registration, and Standards in the Bureau of Radiation Control in the Texas Department of Health. Ms. McBurney was the 2005-2006 president of the Health Physics Society. She has served as treasurer and chairperson of the Conference of Radiation Control Program Directors, and received its highest award, the Gerald S. Parker Award of Merit, in 1994. Ms. McBurney has also served as a representative on the U.S. Nuclear Regulatory Commission's Advisory Committee for the Medical Use of Isotopes and on the U.S. Food and Drug Administration's National Mammography Quality Assurance Advisory Committee. Ms. McBurney received her B.S. in biology from Henderson State University in Arkansas and her M.S. in Radiation Sciences from the University of Arkansas for Medical Sciences. She is a certified health physicist.

Ervin B. Podgorsak is professor and director of the Department of Medical Physics in the Faculty of Medicine at McGill University, in Montreal. Dr. Podgorsak has been at McGill since 1975. His research has ranged from dosimetry to novel irradiation techniques that have resulted in changes in clinical practice. Among his recent research areas is work on stereotactic radiosurgery. Dr. Podgorsak has authored or co-authored 140 peer-reviewed publications, 85 book chapters or conference proceedings, 4 monographs, and 3 textbooks, including a book on radiation physics for medical physicists. He also edited a handbook on radiation oncology physics. He was president of the board of the Canadian College of Physicists in Medicine and was a member of advisory boards for the National Research Council in Ottawa, Canada, the Brookhaven National Laboratory, and the International Atomic Energy Agency. Dr. Podgorsak is a fellow of the Canadian College of Physicists in Medicine and the American Board of Medical Physics. In 2005, he received the Lifetime Achievement Award in Medical Physics from the Upstate New York Chapter of the American Association of Physicists in Medicine (AAPM) and in 2006 the William D. Coolidge Award from the AAPM. Dr. Podgorsak received his B.Sc. in physics from the University of Ljubljana, in Slovenia, and his M.Sc. and Ph.D. in physics from the University of Wisconsin at Madison.

Tor Raubenheimer is an associate professor at the Stanford Linear Accelerator Center (SLAC). He is an expert in accelerator physics, design issues in next generation linear colliders, ion/beam-plasma instabilities in rings and linacs, and effects during bunch length compression. Dr. Raubenheimer has been a Panofsky Fellow at SLAC and a visiting associate scientist at CERN. He received the American Physical Society's Division of Beam Physics Dissertation Award and the U.S. Particle Accelerator School Prize for Achievement in Accelerator Physics and Technology, and is a fellow of the American Physical Society. Dr. Raubenheimer received his B.S. in physics and computer science from Dartmouth College and his Ph.D. in applied physics from Stanford University.

Stephen Wagner is the director of pathogen management and blood product improvement in the Blood Components Department of the American Red Cross's Holland Laboratory for the Biomedical Sciences in Rockville, Maryland. Dr. Wagner received the Daymon Runyon-Walter Winchell Fellowship Grant from the National Cancer Institute and the National Tiffany Award from the American Red Cross for outstanding technical achievement. Dr. Wagner has taught at Bowling Green State University and is a member of the American Society for Microbiology, the American Society for Photobiology, the American Association of Blood Banks, the International Society for Blood Transfusion, and the American Society for Hematology. Dr. Wagner received his B.S. in chemistry from the University of Maryland, College Park, and his M.S. and Ph.D. in biophysics from the Pennsylvania State University.

David L. Weimer is a professor of political science and public affairs at the University of Wisconsin at Madison. Before his appointment at Madison in 2000, he was on the faculty of the University of Rochester. He is an expert in cost-benefit analysis, health policy, and design and functioning of institutions, and also conducts research on energy security, natural resource policy, and education. Dr. Weimer is a past president of the Association for Public Policy Analysis and Management. He has served as a visiting professor at Lingnan College in Hong Kong and as an economist in the Department of Energy's Office of Policy, Planning, and Analysis. He previously served on a committee for the Energy Engineering Board of the National Research Council. Dr. Weimer received his B.S. degree in engineering and applied science and his B.A. in urban studies from the University of Rochester, and received his M.A. in statistics and his M.P.P. and Ph.D. in public policy from the University of California at Berkeley.

APPENDIX B

BACKGROUND ON THE ATOM, RADIOACTIVE DECAY, RADIATION, AND RADIATION DOSE DEPOSITION

This appendix provides introductory and reference information for readers who need background on the structure of atoms, radioactive decay, and forms of radiation and their deposition of energy in materials. This is not meant to be a comprehensive treatment, but should, along with the Glossary in Appendix C, provide the information a reader needs to understand scientific discussions in this report.

ATOMIC AND NUCLEAR STRUCTURE

In the Rutherford model of the atom (Figure B-1), nearly all of the mass and positive charge of the atom are concentrated in the nucleus, the size of which is of the order of 10^{-15} m, and the negative charge is distributed in a cloud outside the nucleus, with a radius of the order of 10^{-10} m. The constituent particles forming an atom are *protons*, *neutrons,* and *electrons*. Protons and neutrons are known as *nucleons* and form the nucleus of the atom. The proton has positive charge and the neutron has no charge. The electron carries a negative charge identical in magnitude to the positive proton charge. The proton and neutron have nearly identical rest masses; the rest mass of the electron is about 2,000 times smaller than that of the proton or neutron.

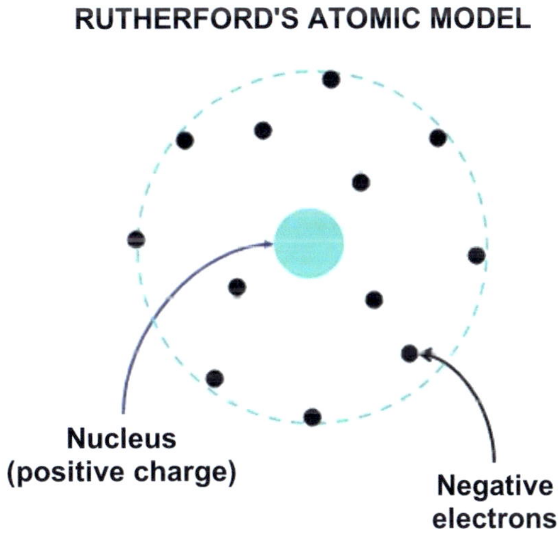

FIGURE B-1 Schematic diagram of the Rutherford atomic model. SOURCE: With kind permission of Springer Science+Business Media.

When discussing topics that involve atomic and nuclear phenomena, it is useful to know the conventions for referring to features of atoms and nuclei, including the following.

- *Atomic number* Z is the number of protons and number of electrons in an atom.
- *Atomic mass number* A is the number of nucleons in an atom, that is, number of protons Z plus number of neutrons N in an atom; $A = Z + N$.
- *Atomic mass* $\mathcal{M}$ is expressed in atomic mass units u, where 1 u is equal to 1/12th of the mass of the carbon-12 atom. The atomic mass $\mathcal{M}$ is smaller than the sum of individual masses of constituent particles because of the intrinsic energy associated with binding the particles (nucleons) within the nucleus.

In nuclear physics, a nucleus X is designated as ${}^{A}_{Z}X$, where A is the atomic mass number and Z the atomic number. For example, the cobalt-60 nucleus is identified as ${}^{60}_{27}Co$; the radium-226 nucleus as ${}^{226}_{88}Ra$. Because both Z and the chemical symbol uniquely identify the element, Z is commonly omitted leaving ${}^{60}Co$ and ${}^{226}Ra$.

An element may be composed of atoms that all have the same number of protons, that is, have the same atomic number Z, but have different numbers of neutrons, that is, have different atomic mass numbers A. Such atoms of identical atomic number Z but differing atomic mass numbers A are called *isotopes* of a given element.

The term isotope is often misused to designate nuclear species. For example, cobalt-60, cesium-137, and radium-226 are not isotopes, since they do not belong to the same element. Rather than isotopes, they should be referred to as *nuclides*. On the other hand, it is correct to state that deuterium (with nucleus called deuteron) and tritium (with nucleus called triton) are heavy isotopes of hydrogen or that cobalt-59 and cobalt-60 are isotopes of cobalt. The term *radionuclide* should be used to designate radioactive species; however, the term radioisotope is often used for this purpose.

If a nucleus exists in an excited state for some time, it is said to be in an isomeric (metastable) state. *Isomers* thus are nuclear species that have common atomic number Z and atomic mass number A. For example, technetium-99*m* is an isomeric state of technetium-99 and cobalt-60*m* is an isomeric state of cobalt-60.

RADIOACTIVE DECAY

Henri Becquerel discovered natural radioactivity in 1896; Pierre and Marie Curie discovered radium in 1898. These basic discoveries stimulated subsequent discoveries, such as artificial radioactivity by Frédéric and Irène Joliot-Curie in 1934 and neutron-induced fission by Otto Hahn, Fritz Strassmann, Lise Meitner, and Otto Frisch in 1939. All these discoveries are of tremendous importance in science, medicine, and industry.

Radioactivity

Radioactivity is a process by which an unstable parent nucleus transforms spontaneously into one or several daughter nuclei. These are more stable than the parent nucleus but may still be unstable and will decay further through a chain of radioactive decays until a stable nuclear configuration is reached. Radioactive decay is usually accompanied by emission of energetic particles and/or gamma rays which together form a class of radiation that is referred to as ionizing radiation.

- *Nuclear decay*, also called *nuclear disintegration*, *nuclear transformation,* or *radioactive decay*, is a statistical phenomenon.
- The exponential laws that govern nuclear decay and growth of radioactive substances were first formulated by *Ernest Rutherford* and *Frederick Soddy* in 1902 and then refined by *Harry Bateman* in 1910.
- A radioactive substance containing atoms of the same structure is referred to as a radioactive nuclide. Radioactive atoms, like any other atomic structure, are characterized by the atomic number *Z* and atomic mass number *A.*
- Radioactive decay involves a transition from the quantum state of the original nuclide (parent) to a quantum state of the product nuclide (daughter). The energy difference between the two quantum levels involved in a radioactive transition is referred to as the decay energy *Q*. The decay energy is emitted either in the form of electromagnetic radiation (usually gamma rays) or in the form of kinetic energy of the reaction products.
- The mode of radioactive decay depends upon the particular nuclide involved.
- Radioactive decay processes are governed by general formalism that is based on the definition of the activity $\mathcal{A}(t)$ and on a characteristic parameter for each radioactive decay process: radioactive *decay constant* λ with dimensions of reciprocal time usually in s^{-1}.
- The *radioactive decay constant* λ multiplied by a time interval that is much smaller than $1/\lambda$ represents the probability that any particular atom of a radioactive substance containing a large number $N(t)$ of identical radioactive atoms will decay (disintegrate) in that time interval. An assumption is made that λ is independent of the physical environment of a given atom.
- *Activity* $\mathcal{A}(t)$ of a radioactive substance containing a large number $N(t)$ of identical radioactive atoms represents the total number of decays (disintegrations) per unit time and is defined as a product between $N(t)$ and λ.
- The *SI unit of activity* is the becquerel (Bq), given as $1\,\text{Bq} = 1\,\text{s}^{-1}$. The becquerel and hertz both correspond to s^{-1}, but hertz expresses frequency of periodic motion, while becquerel expresses activity.
- The old unit of activity, the curie (Ci), was initially defined as the activity of 1 g of radium-226 and given as $1\,\text{Ci} = 3.7 \times 10^{10}\ \text{s}^{-1}$. The activity of 1 g of radium-226 was subsequently measured to be $3.665 \times 10^{10}\ \text{s}^{-1}$; however, the definition of the curie was kept at $3.7 \times 10^{10}\ \text{s}^{-1}$. The current value of the activity of 1 g of radium-226 is thus 0.988 Ci or $3.665 \times 10^{10}\ \text{Bq}$.
- Bq and Ci are related as follows: $1\,\text{Bq} = 2.703 \times 10^{-11}$ Ci or $1\,\text{Ci} = 3.7 \times 10^{10}\ \text{Bq}$.
- The specific activity *a* is defined as activity $\mathcal{A}$ per unit mass M. The specific activity of a radioactive nuclide depends on the decay constant λ and on the atomic mass number *A* of the radioactive nuclide. The units of specific activity are Bq/kg (SI unit) and Ci/g (old unit).
- The half-life of a radioactive substance is that time during which the number of radioactive nuclei of the substance decays to half of the initial value. One may also state that in the time equal to one half-life of a radionuclide the activity of the radionuclide diminishes to one-half of its initial value.

Radionuclides

Because they produce ionizing radiation through radioactive decay, radionuclides play an important role in science, industry, and medicine, but are also of concern to humans because, when used unsafely or with malicious intent, they may have significant deleterious effects on human tissues. These effects depend on the type of tissue and the dose absorbed by the tissue. Radionuclides can irradiate humans as a source of external radiation (radionuclide located outside but in the vicinity of the human body) or a source of internal radiation (radionuclide enters the body through various pathways such as ingestion, inhalation, or through skin).

Radioactive nuclides (radionuclides) are divided into two categories: naturally occurring and man-made or artificially produced. Aside from their origins, there is no essential physical difference between the two categories of radionuclides and the division is mainly historical.

- The man-made (artificial) radionuclides are manufactured by bombarding stable or very-long-lived nuclides with energetic particles produced by machines of various kinds, such as nuclear reactors, cyclotrons, and linear accelerators. The process of radionuclide production is referred to as radioactivation or nucleosynthesis. Currently, the list of known nuclides contains some 275 stable nuclides and over 3,000 radionuclides.
- Many of the known radionuclides used in industrial and medical applications are produced artificially through radioactivation; however, there are also radionuclides created through fission (splitting) of heavy nuclei and a few naturally occurring radionuclides, almost exclusively members of one of four natural radioactive series that all begin with very-long-lived parents with half-lives comparable to the age of the Earth. The long-lived parents decay through several radioactive daughter products, eventually to reach a stable lead nuclide or a stable bismuth-203 nuclide. Most notable other examples of naturally occurring radionuclides are carbon-14, produced by cosmic ray protons, and the long-lived potassium-40, which occurs naturally.

Radionuclides are unstable and strive to reach more stable nuclear configurations through various processes of spontaneous radioactive decay. General aspects of spontaneous radioactive decay may be discussed using the formalism based on the definitions of activity and decay constant without regard to the actual microscopic processes that underlie the radioactive disintegrations. In each nuclear transformation a number of physical quantities must be conserved. The most important of these quantities are total energy, momentum, charge, atomic number, and atomic mass number.

Modes of Radioactive Decay

A closer look at radioactive decay processes shows that they are divided into seven main categories:

1. alpha decay producing alpha particles,
2. beta minus decay producing negative beta particles (electrons),
3. beta plus decay producing positive beta particles (positrons),
4. electron capture,
5. gamma decay producing gamma rays,
6. internal conversion producing energetic electrons,
7. spontaneous fission producing neutrons and fission fragments.

The seven modes of radioactive decay are characterized by their own decay mechanism and constraints imposed by conservation laws; however, they all follow the same statistical process described by a simple exponential function. The main characteristics of the seven decay modes are as follows:

- **Alpha decay** was the first mode of radioactive decay detected and investigated during the 1890s. It played a very important role in early modern physics experiments that led to the Rutherford atomic model and is characterized by a nuclear transformation in which an unstable parent nucleus attains a more stable nuclear configuration through ejection of an alpha particle (nucleus of helium-4 atom) with kinetic energy of the order of a few million electron volts (MeV). Alpha emitters pose no danger to humans as external sources because the alpha particles have a short range in air (of the order of a few centimeters) and cannot penetrate the superficial dead layer of the skin. However, when ingested or inhaled, alpha emitters are dangerous because they interact with living tissues such as bone marrow or alveoli in the lung, potentially depositing a very high radiation dose to internal human tissues.
- **Beta minus decay** is characterized by a nuclear transformation in which a neutron transforms into a proton, and an electron and anti-neutrino, sharing the available energy, are ejected from the nucleus. Energetic electrons emitted in beta minus decay have a relatively small mass and can penetrate human tissue to a depth of a few centimeters, and so they pose a hazard to humans both as external and internal radiation sources.
- **Beta plus decay** is characterized by a nuclear transformation in which a proton transforms into a neutron, and a positron and neutrino, sharing the available energy, are ejected from the nucleus. Energetic positrons emitted in beta plus decay have a relatively small mass and can penetrate human tissue to a depth of a few centimeters, and so they pose a hazard to humans both as external and internal radiation sources. None of the radionuclides considered in this report decay by beta plus decay.
- In **electron capture** the nucleus captures one of its own atomic shell electrons, a nuclear proton transforms into a neutron, and a neutrino is ejected. The process competes with the beta plus decay.
- **Gamma decay** results from a transition between nuclear excited states or a transition from an excited state to the ground state of a nucleus. The nucleus does not undergo a transformation, but nuclear transitions are typically accompanied by emission of gamma rays with energies of the order of 1 MeV. These gamma rays can penetrate deep into the human body and are thus hazardous to humans both as external or internal sources. The damage that they do to cells and tissue can be used for beneficial purposes, as in the case of radiotherapy to treat malignant tumors.
- The energy available for a gamma-ray transition may be transferred to an atomic electron, which is ejected with a relatively large kinetic energy. The process is referred to as **internal conversion** and competes with gamma decay.
- In addition to decaying through alpha and beta decay processes, nuclei with very large atomic mass numbers *A* may also disintegrate by splitting into two nearly equal fission fragments and concurrently emit two to four neutrons. This decay process, called **spontaneous fission**, competes with alpha decay and is accompanied by liberation of a significant amount of energy. It is similar to the standard nuclear fission process except that it is not self-sustaining, since it does

not generate the neutron fluence rate required to sustain a chain reaction. For practical purposes, the most important radionuclide undergoing the spontaneous fission decay is californium-252, used in industry and brachytherapy as an efficient source of fast neutrons.

Radioactive Decay of Key Radionuclides Used in industry and Medicine

Of the 3,000 known radionuclides, about 200 are used in industry and medicine. For the purposes of this report, four radionuclides are of special interest as a result of their widespread use in industry and medicine. The four radionuclides are cobalt-60, cesium-137, iridium-192, and americium-241; diagrams that illustrate the decay modes and energies of these radionuclides appear in Figures B-2 to B-5, respectively.

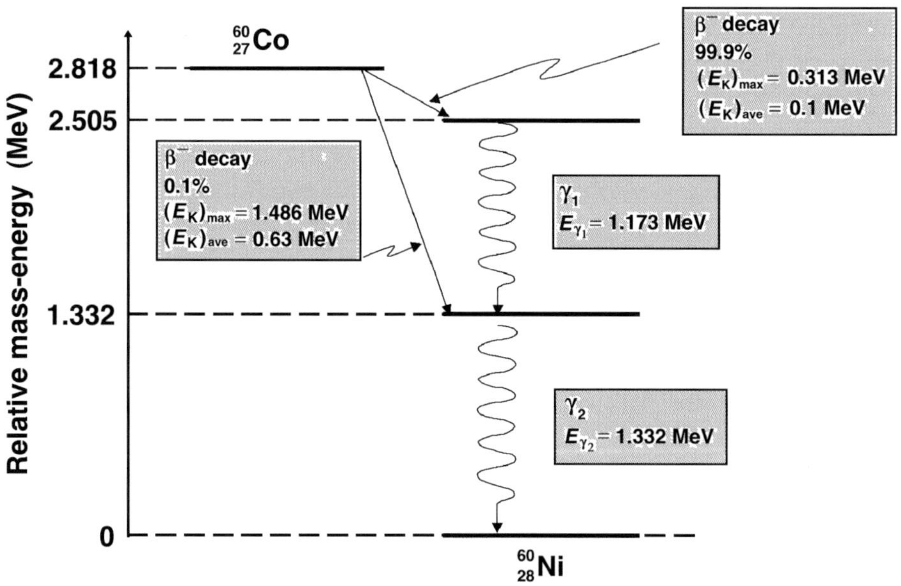

FIGURE B-2 Decay scheme for the beta minus decay of cobalt-60 into nickel-60 with a half-life of 5.26 years. The cobalt-60 nucleus transforms into a nickel-60 nucleus in the second excited state (99.9% of disintegrations). The excited nickel-60 nucleus decays instantaneously from the second excited state into the first excited state by emitting a 1.17-MeV gamma photon, and from the first excited state to the ground state by emitting a 1.33-MeV gamma photon. Note: the two gamma photons are called cobalt-60 gamma rays, yet they actually originate in nickel-60. The average energy of the two is 1.25 MeV. SOURCE: With kind permission of Springer Science+Business Media.

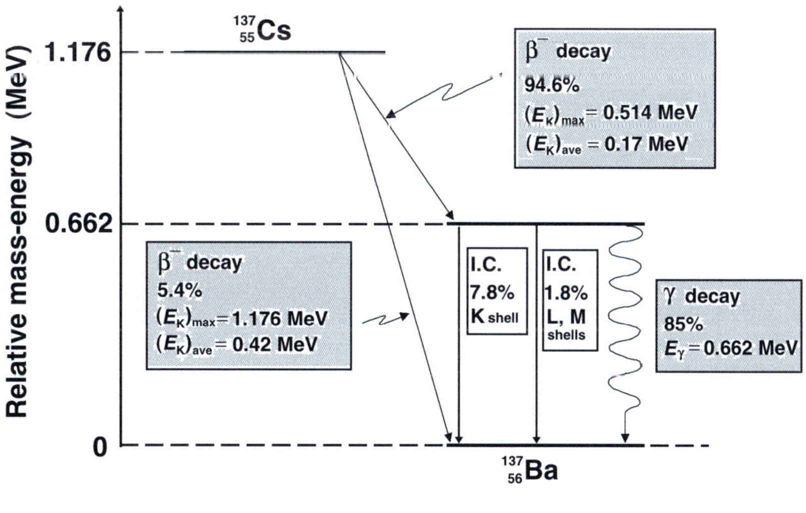

FIGURE B-3 Decay scheme for the beta minus decay of cesium-137 into barium-137 with a half-life of 30 years. The cesium-137 nucleus transforms into a barium-137 nucleus in the first excited state (94.6% of all disintegrations). The excited barium nucleus decays into its ground state by emitting a 0.662-MeV gamma photon (85% of disintegrations). Note: The 0.662-MeV gamma photon is called a cesium-137 gamma photon, yet it is emitted by the barium-137 nucleus. SOURCE: With kind permission of Springer Science+Business Media.

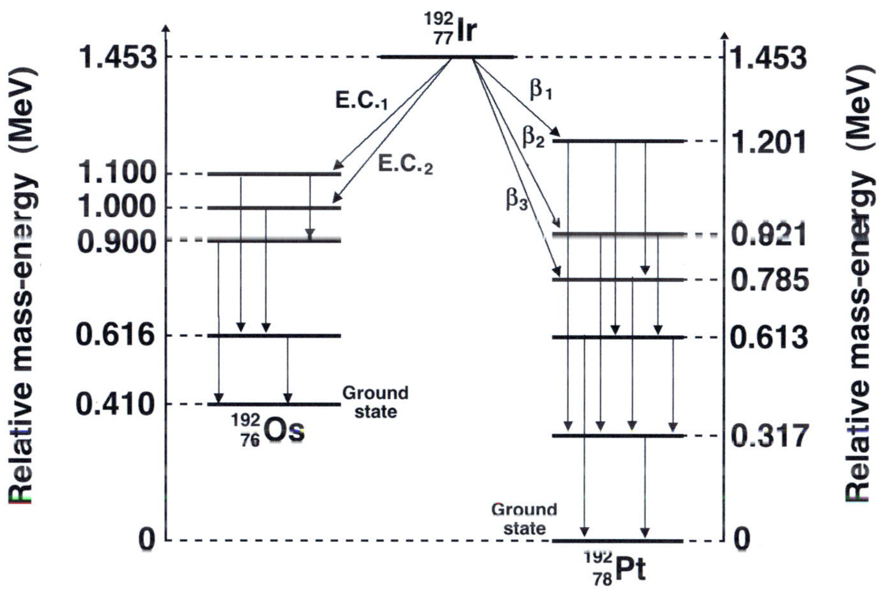

FIGURE B-4 Decay scheme for the beta minus decay of iridium-192 into platinum-192 and the electron capture decay of iridium-192 into osmium-192 with a half-life of 74 days. Both platinum-192 and osmium-192 are produced in various excited states and they both instantaneously attain their ground states through emission of gamma photons. The spectrum of iridium-192 gamma photons thus consists of many gamma photons, and the effective energy of the iridium-192 gamma photons is of the order of 0.34 MeV. Note: The gamma photons emitted by an iridium-192 source are called iridium-192 gamma photons, yet they are emitted by either osmium-192 or platinum-192. SOURCE: With kind permission of Springer Science+Business Media.

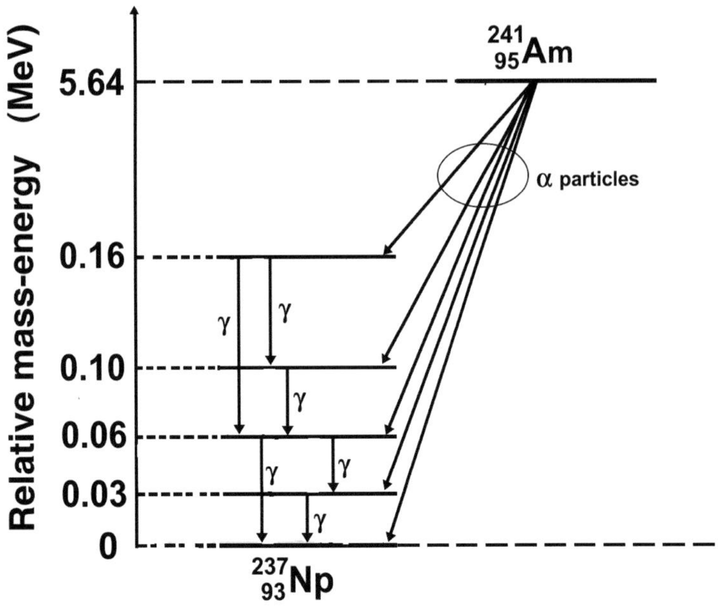

FIGURE B-5 Decay scheme for the alpha decay of americium-241 into neptunium-237 with a half-life of 432 years. A spectrum consisting of five discrete alpha particle energies is emitted and the alpha particles produce neptunium-237 in four excited states in addition to the ground state. All excited states instantaneously decay through various excited states into the ground state of neptunium-237 through emission of gamma photons. SOURCE: Image provided by committee.

FORMS OF RADIATION

Radiation is referred to as non-ionizing or ionizing (see Figure B-6), depending on its ability to ionize matter.

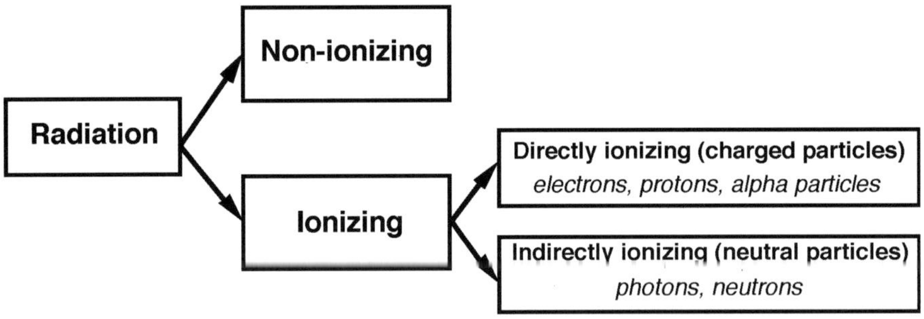

FIGURE B-6 Classification of radiation. SOURCE: With kind permission of Springer Science+Business Media.

As the term implies, **ionizing radiation**, in contrast to non-ionizing radiation, is characterized by its ability to ionize atoms and molecules of matter, thereby producing ions and energetic electrons. These ionizing processes have many useful purposes in medicine and industry, but can also cause serious unwanted biological damage in living tissues when used carelessly or with malicious intent. Energy is transferred from ionizing radiation to an absorbing medium through Coulomb interactions with orbital electrons and nuclei constituting the atoms and molecules of the absorbing medium. Ionizing radiation is split into two main categories: directly ionizing and indirectly ionizing.

Directly Ionizing Radiation

Directly ionizing radiation consists of charged subatomic particles such as light charged particles (electrons and positrons) and heavy charged particles (protons, alpha particles, and heavier ions) which experience direct Coulomb interactions with orbital electrons and nuclei of the absorbing medium. As the term implies, energy is transferred from the charged particle to atomic orbital electrons in a direct manner.

It originates from:

- radioactive decay producing alpha particles in alpha decay, beta particles in beta decay, electrons in internal conversion;
- a particle accelerator producing energetic electrons in linear accelerator, betatron, microtron, or synchrotron;
- a particle accelerator producing energetic protons or heavier ions in cyclotron, synchrocyclotron, or synchrotron.

Indirectly Ionizing Radiation

Indirectly ionizing radiation consists of energetic neutral "particles" such as x rays, gamma rays, and neutrons. These neutral particles first transfer energy to energetic charged particles and these charged particles, as they move through the absorbing medium, experience Coulomb interactions with orbital electrons and nuclei of the absorbing medium. Energy transfer from neutral particles to absorbing medium is thus a two-step process, hence the term indirect ionization. In the case of x-rays and gamma rays, the energetic charged particles released in the absorbing medium are electrons or positrons produced through photoelectric effect, Compton effect or pair production; in the case of neutrons, these energetic charged particles are protons or heavier nuclei released in the absorbing medium through nuclear reactions.

It originates from:

- radioactive decay producing gamma rays in gamma decay, neutrons in spontaneous fission;
- an electron accelerator producing x-rays in x-ray machine, linear accelerator, betatron, or microtron;
- a neutron generator producing neutrons in a charged particle electrostatic accelerator or through bombarding a beryllium target with alpha particles produced through alpha decay (e.g., Am-Be).

Indirectly ionizing photon radiation consists of four distinct groups of photons:

- *characteristic (fluorescent) x rays* which result from electron transitions between atomic shells;
- *bremsstrahlung x rays* which result from Coulomb interactions between an electron and atomic nucleus;
- *gamma rays* which result from nuclear transitions;
- *annihilation quanta* which result from positron-electron annihilation.

Indirectly ionizing neutron radiation is classified according to neutron kinetic energy (E_k):

- *ultracold neutrons with* $E_k < 2 \times 10^{-7}$eV,
- *very cold neutrons with* 2×10^{-7} eV $\leq E_k \leq 5 \times 10^{-5}$ eV,
- *cold neutrons with* 5×10^{-5} eV $\leq E_k \leq 0.025$ eV,
- *thermal neutrons* with $E_k \approx 0.025$ eV,
- *epithermal neutrons* with 1 eV $\leq E_k \leq$ 1 keV,
- *intermediate neutrons* with1 keV $\leq E_k \leq$ 0.1 MeV,
- *fast neutrons* with $E_k > 0.1$ MeV.

X-rays

The importance of ionizing radiation was recognized astonishingly quickly after Wilhelm Roentgen discovered X-rays on November 8, 1895. On December 22 of that year, Dr. Roentgen made a now famous image of his wife Bertha's hand showing her bones and her wedding ring (see Figure B-7). By January 16, 1896, news and understanding of the momentous nature of the discovery was reported in the *New York Times*, which predicted the "transformation of modern surgery by enabling the surgeon to detect the presence of foreign bodies;" and indeed battlefield physicians began using x rays to find bullets in wounded soldiers and fractures in their bones within months (Asmuss, 1995).

The clinical use of x-rays spread rapidly across North America immediately after the discovery of X-rays in 1895 starting in the United States with Yale University on January 27, 1896, and Dartmouth College on February 3, 1896, and in Canada with McGill University on February 5, 1896, and the University of Toronto on February 7, 1896.

During the first two decades after 1895, x-rays were produced with low-pressure glass tubes incorporating two electrodes, referred to as the Crookes tube. In the Crookes tube (see Figure B-8a) the potential difference between the two electrodes produces discharge in the rarified gas, causing ionization of gas molecules. Electrons (cathode rays) are accelerated toward the positive electrode, producing x-rays upon striking it. In 1913, William D. Coolidge introduced his invention of ductile tungsten into the x-ray tube and revolutionized x-ray tube design (see Figure B-8b) with the use of a hot-filament cathode for the source of electrons in high-vacuum tubes. Hot cathodes emit electrons through thermionic emission and are still in use today in modern x-ray tubes, now called Coolidge tubes, and in electron guns of linear accelerators. The maximum x-ray energy produced in the x-ray target (anode) equals the kinetic energy of electrons striking the target.

X-ray radiographs have advanced to a point where detailed imaging of an entire human body can be carried out in clinical practice. A computed tomography (CT) scanner is a machine that uses an x-ray beam rotating about a specific area of a patient to collect x-ray attenuation data for the patient's tissues. It then manipulates these data with special mathematical algorithms to display a series of transverse slices through the patient. The transverse CT data can be reconstructed so as to obtain coronal and sagittal section through the patient's organs or to obtain digitally reconstructed radiographs. The excellent resolution obtained with a modern CT scanner provides an extremely versatile "non-invasive" diagnostic tool. CT scanners have been in clinical and industrial use since the early 1970s and evolved through five generations, each generation increasingly more sophisticated and faster.

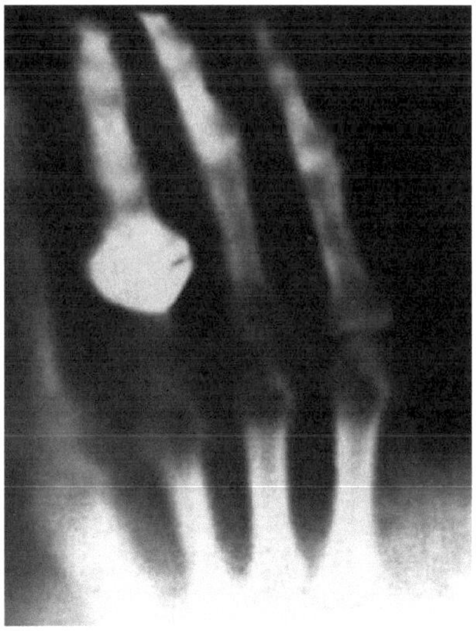

FIGURE B-7 A radiograph of Bertha Roentgen's hand taken in 1895 as the first example of the potential that Roentgen's discovery had for clinical use.

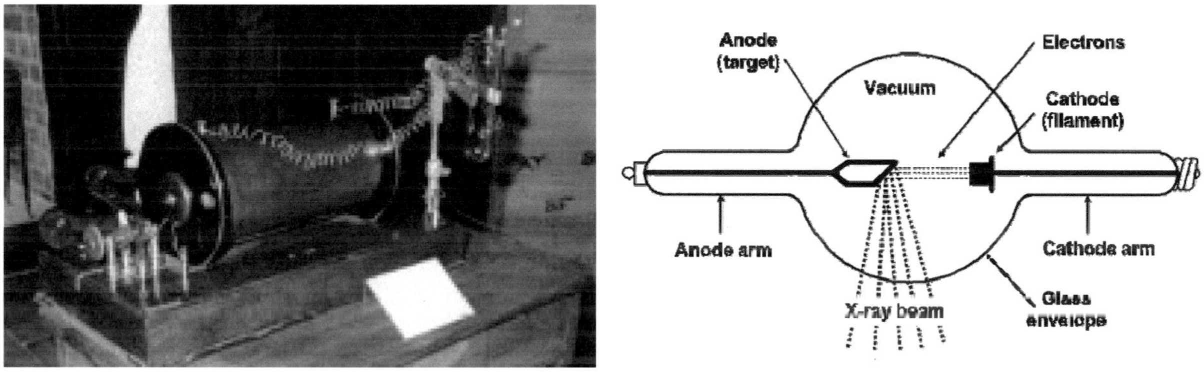

FIGURE B-8 (a) Photograph of Roentgen's apparatus for production of x-rays, referred to as Crookes cold cathode tube; (b) schematic diagram of a Coolidge tube, referred to as hot cathode tube.

DOSE DEPOSITION IN WATER FOR VARIOUS IONIZING RADIATION BEAMS

In benign use of ionizing radiation, the dose deposition in matter is one of the most important characteristics defining the effectiveness of a particular usage. For example, in medical physics the dose deposition properties in water and tissue govern the diagnosis of disease with radiation (*imaging physics*), treatment of disease with radiation (*radiation oncology physics*), and the study of deleterious effects of ionizing radiation on humans (*health physics*). Food irradiation, sterilization of medical equipment with ionizing radiation, and blood irradiation depend heavily on the delivered absorbed dose in the irradiated substances for the successful outcome of the irradiation procedure. In the use of ionizing radiation in industrial imaging, it is the interactions of the radiation with the imaged objects and the dose deposition in the image receptors that govern the image quality as well as the radiation safety requirements.

Medical imaging with ionizing radiation is limited to the use of x-ray beams in *diagnostic radiology* and gamma rays in *nuclear medicine* (see Figure B-9a), while in *radiation oncology* (see Figure B-9b) the use of ionizing radiation is broader and covers essentially all ionizing radiation types ranging from x rays and gamma rays through electrons to neutrons, protons, and heavier charged particles. In diagnostic radiology and industrial radiography, one is interested in the radiation beam that propagates through the patient or imaged object, respectively, while in nuclear medicine, one is interested in the radiation that emanates from the patient. In radiation oncology, on the other hand, one is interested in the energy deposited in the patient (radiation dose) by a radiation source that is located outside the patient (external beam radiotherapy) or placed directly inside the tumor (brachytherapy).

When considering the dose deposition in tissue by radiation beams, four beam categories are usually defined: two categories (*photons* and *neutrons*) for indirectly ionizing radiation and two categories (*electrons* and *heavy charged particles*) for directly ionizing radiation. Since water is the main constituent of human tissue, it is for practical reasons often used as tissue substitute in the determination of the interaction of ionizing radiation with human tissue and the absorbed dose in tissue.

Figure B-10 displays depth doses in water or patient normalized to 100 percent at the depth of dose maximum (percent depth doses) for various ionizing radiation types and energies: for indirectly ionizing radiation (in [a] for *photons* and in [b] for *neutrons*) and for directly ionizing radiation (in [c] for *electrons* and in [d] for *protons*). It is evident that the depth dose characteristics of radiation beams depend strongly upon beam type and energy. However, they also depend in a complex fashion on other beam parameters, such as field size and source-patient distance. In general, indirectly ionizing radiation exhibits exponential-like attenuation in absorbing media, whereas directly ionizing radiation exhibits a defined range in absorbing media. Although all four radiation categories are used in radiotherapy, routine radiotherapy is generally done with x-rays, gamma rays, or electrons, the beam choice depending on the location of the treated tumor and availability of equipment.

Dose Distributions for Photon Beams in Water

A photon beam propagating through air or vacuum is governed by the inverse-square law which results in the diminution of the beam's intensity as the inverse square of the distance from the radiation source. A photon beam propagating through water or a patient, on the other hand, is not only affected by the inverse-square law, but also by attenuation and scattering of the photon beam inside the patient. These three effects make the dose deposition in a patient a complicated process and its determination a complex task.

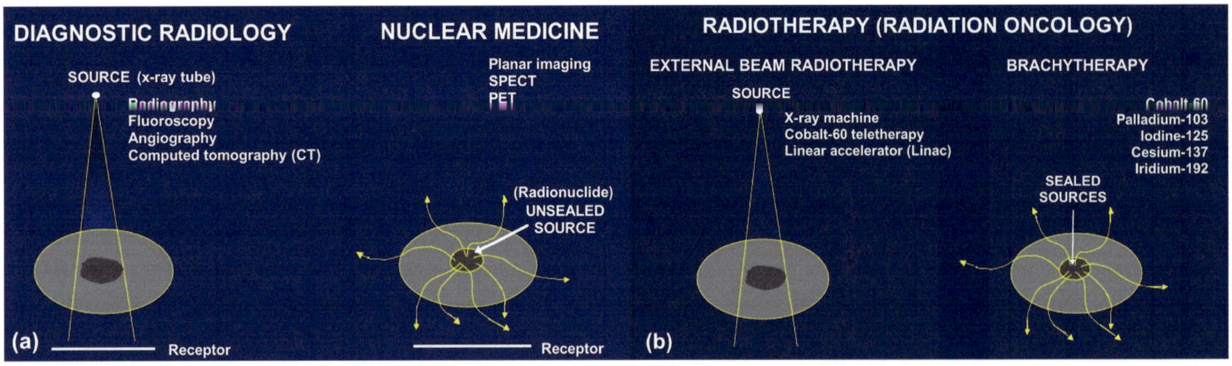

FIGURE B-9 Schematic diagram of ionizing radiation use in medicine for (a) diagnosis of disease (diagnostic radiology and nuclear medicine) and (b) treatment of disease (external-beam radiotherapy and brachytherapy). SOURCE: Provided by the committee.

Typical dose distributions in water for several photon beams in the energy range from 100 kVp to 22 MV are shown in Figure B-10a). Several important points and regions of the absorbed dose curves may be identified. The beam enters the patient on the surface, where it delivers a certain surface dose D_s. Beneath the surface the dose first rises rapidly with depth, reaches a maximum value at a depth z_{max}, and then decreases almost exponentially until it reaches an exit dose value at the patient's exit point.

The depth of dose maximum is proportional to the beam energy and typically amounts to 0 for superficial (50–100 kVp) and orthovoltage (100–300 kVp) beams; 0.5 cm for cobalt-60 gamma rays; 1.5 cm for 6-MV beams; 2.5 cm for 10-MV beams; and 4 cm for 22-MV beams.

The relatively low surface dose for high-energy photon beams (referred to as the *skin sparing effect*) is of great importance in radiotherapy for treatment of deep-seated lesions without involvement of the skin. The tumor dose can be concentrated at large depths in the patient concurrently with delivering a low dose to patient's skin that is highly sensitive to radiation and must be spared as much as possible when it is not involved in the disease.

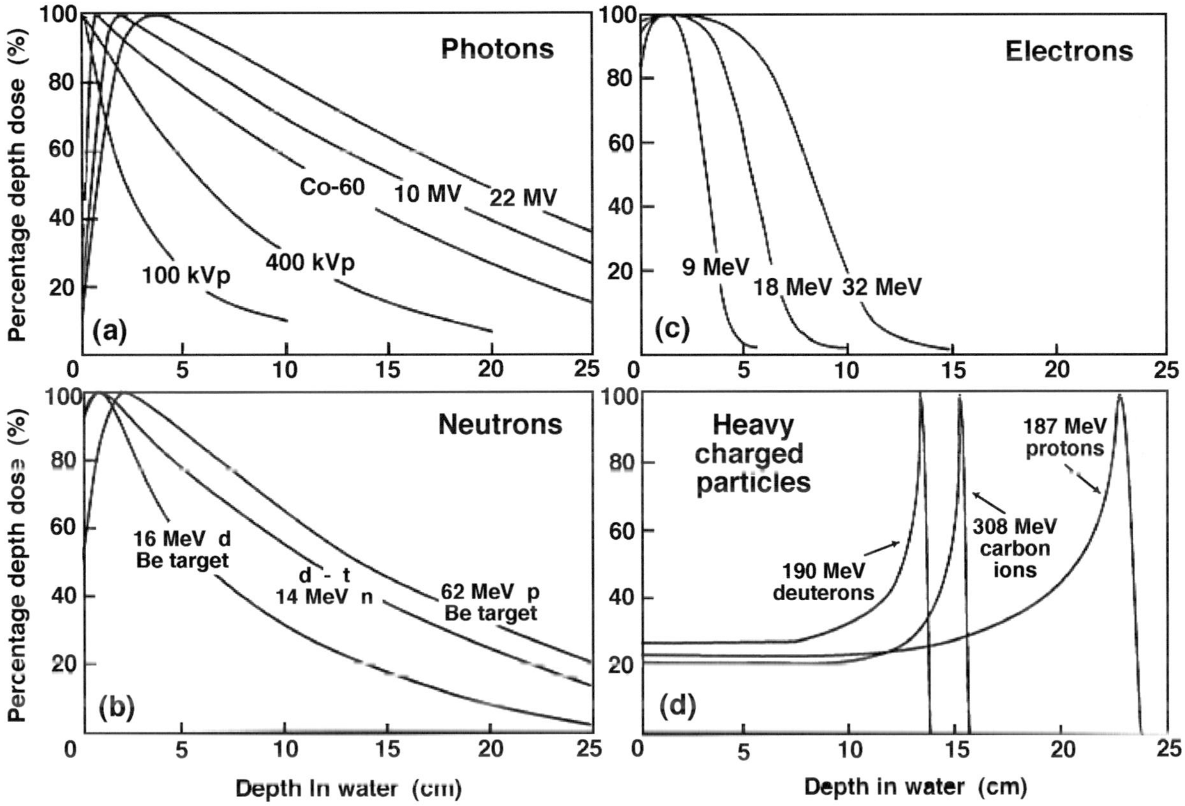

FIGURE B-10 Absorbed dose against depth in water for ionizing radiation beams of various types and energies. Parts (a) and (b) are for *indirectly ionizing radiation*: in (a) for photon beams in the range from 100 kVp to 22 MV and in (b) for neutron beams. Parts (c) and (d) are for *directly ionizing radiation*: in (c) for megavoltage electron beams in the range from 9 to 32 MeV and in (d) for heavy charged particle beams (187-MeV protons, 190-MeV deuterons, and 308-MeV carbon ions). SOURCE: With kind permission of Springer Science+Business Media.

The dose region between the surface and the depth of dose maximum z_{max} is called the dose *buildup region* and represents the region in the patient in which the dose deposition rises with depth as a result of the range of secondary electrons released in tissue by photon interactions with the atoms of tissue. It is these secondary electrons released by photons that deposit energy in tissue (indirect ionization). The larger is the photon energy, the larger is the range of secondary electrons and, consequently, the larger is the depth of dose maximum.

Dose Distributions for Neutron Beams in Water

Neutron beams belong to the group of indirectly ionizing radiation, but rather than releasing electrons like photons do, they release protons or heavier nuclei which then deposit their energy in absorbing media through Coulomb interactions with the electrons and nuclei of the absorber.

As shown in Figure B-10b, the dose deposition characteristics of neutrons in water are similar to those of photon beams. Neutron beams exhibit a relatively low surface dose, although the skin sparing effect is less pronounced than that for energetic photon beams. They also exhibit a dose maximum beneath the skin surface and an almost exponential decrease in dose beyond the depth of dose maximum. The dose buildup region depends on neutron beam energy; the larger is the energy, the larger is the depth of dose maximum.

For comparison, we may state that a 14-MeV neutron beam has depth dose characteristics that are comparable to a cobalt-60 gamma-ray beam; a 65-MeV neutron beam is comparable to a 10-MV x-ray beam.

Dose Distributions for Electron Beams in Water

Electrons are directly ionizing radiation that deposits the energy in tissue through Coulomb interactions with orbital electrons and nuclei of the absorber atoms. Megavoltage electron beams represent an important treatment modality in modern radiotherapy, often providing a unique option for treatment of superficial tumors that are less than 5 cm deep. Electrons have been used in radiotherapy since the early 1950s, first produced by betatrons and then by linear accelerators. Modern high-energy linear accelerators typically provide, in addition to two megavoltage x-ray beams, several electron beams with energies from 4 to 25 MeV.

As shown in Figure B-10c, the electron-beam dose distribution with depth in patient exhibits a relatively high surface dose and then builds up to a maximum dose at a certain depth, referred to as the electron-beam depth dose maximum z_{max}. Beyond z_{max} the dose drops off rapidly, and levels off at a small low-level dose component referred to as the *bremsstrahlung* tail. Several parameters are used to describe clinical electron beams, such as the most probable energy on the patient's skin surface, the mean electron energy on the patient's skin surface, or the depth at which the absorbed dose falls to 50 percent of the maximum dose.

Unlike the case for photon beams, the depth of dose maximum for electron beams does not depend on beam energy; rather it is a function of machine design. On the other hand, the beam penetration into tissue clearly depends on beam energy; the higher is the electron beam energy, the more penetrating is the electron beam, as is evident from Figure B-10c.

The bremsstrahlung component of the electron beam is the photon contamination that results from radiation losses experienced by the incident electrons as they penetrate the various machine components, air, and the patient. The higher is the energy of the incident electrons, the higher is the bremsstrahlung contamination of the electron beam.

Dose Distributions for Heavy Charged Particle Beams in Water

Heavy charged particle beams fall into the category of directly ionizing radiation and deposit their energy in tissue through Coulomb interactions with orbital electrons of the absorber. As they penetrate into tissue, heavy charged particles lose energy but, in contrast to electrons, do not diverge appreciably from their direction of motion and therefore exhibit a distinct range in tissue. This range depends on the incident particle's kinetic energy and mass.

Just before the heavy charged particle expends all of its kinetic energy, its energy loss per unit distance traveled increases drastically and this results in a high dose deposition at that point in tissue. As shown in Figure B-10d, this high-dose region appears close to the particle's range, is very narrow, and defines the maximum dose deposited in tissue. This peak dose is referred to as the Bragg peak, and it characterizes all heavy charged particle dose distributions.

Because of their large mass compared to the electron mass, heavy charged particles lose their kinetic energy, only interacting with orbital electrons of the absorber. Since they do not lose any appreciable amount of energy through bremsstrahlung interactions with absorber nuclei, their depth dose curves do not exhibit a bremsstrahlung contamination tail.

APPENDIX C

GLOSSARY

10 CFR Part 61.55: Title 10, Part 61, Section 55 of the *Code of Federal Regulations*, Waste classification. These regulations were promulgated by the U.S. Nuclear Regulatory Commission.

A/D ratio: The ratio of the activity (A) of a radiation source to the activity determined to define a threshold level of danger (D) in the International Atomic Energy Agency's radiation source categorization system.

Absorbed dose: The quantity of ionizing radiation deposited into a material, including an organ or tissue, expressed in terms of the energy absorbed per unit mass of material. The basic unit of absorbed dose is the rad or its SI equivalent, the gray (Gy).

Accelerator: A device that accelerates charged subatomic particles. Also called a particle accelerator, in the context of this report these devices are used to generate energetic beams of electrons that can be directed at an object that one wants to irradiate or at a tungsten, tantalum, or gold target, which converts the electron energy into x-rays that irradiate an object.

Actinide: Any of a series of chemically similar radioactive elements with atomic numbers ranging from 89 (actinium) through 103 (lawrencium). This group includes uranium (atomic number 92), plutonium (atomic number 94), and americium (atomic number 95).

Activity: The rate of decay of a radionuclide, more formally, the number of decays per time. Its SI unit is the becquerel (Bq) corresponding to one radioactive decay (disintegration) per second; its old unit, the curie (Ci), was originally defined as the activity of 1 gram of radium-226 or 3.7×10^{10} disintegrations per second.

Acute effect: Effects in organisms manifest themselves soon after exposure to radiation and are characterized by inflammation, edema, denudation and depletion of epithelial and haemopoietic tissue, and haemorrhage.

Acute radiation exposure: A radiation exposure that occurs over a relatively short period of time (e.g., seconds to hours). A chest X-ray is an acute radiation exposure.

Agreement State: States that have assumed authority under Section 274b of the Atomic Energy Act to license and regulate by-product materials (radioisotopes), source materials (uranium and thorium), and certain quantities of special nuclear materials.

Area-denial RDD: A radiological dispersal device (RDD) intended to contaminate an area such that the area cannot be occupied or used.

Atomic number: The number of protons in the nucleus of an atom, and the number of electrons in a neutral atom. This number, sometimes referred to by the symbol *Z*, determines the atom's chemical element. A high-*Z* material has a high atomic number.

Becquerel (Bq): A unit of measure for activity. One becquerel is 1 disintegration (radioactive decay) per second. A gigabecquerel (GBq) is 10^9 Bq (1 billion becquerels) and a terabecquerel (TBq) is 10^{12} Bq (1 million million becquerels).

Bremsstrahlung: Radiation emitted by the slowing down of light charged particles, such as the x rays produced when electrons from an accelerator are stopped in a metal target.

By-product material: Defined by the Atomic Energy Act as radioactive material (except special nuclear material) yielded in or made radioactive by exposure to the radiation incident to the process of producing or using special nuclear material; and tailings or wastes produced by the extraction or concentration of uranium or thorium from any ore processed primarily for its source material content.

Cancer incidence: Also known as the *incidence rate*. The rate of occurrence of cancer within a specified period of time per unit of population, for example, the number of cancers per year per 100,000 people.

Cancer mortality: Also known as the *mortality rate*. The rate of death from cancer within a specified period of time per unit of population, for example, number of cancer deaths per year per 100,000 people.

Carcinogenesis (induction of cancer): Manifested as a late stochastic somatic effect in the form of acute or chronic myeloid leukemia or some solid tumors, for example, in the skin, bone, lung, thyroid, or breast.

Category 1 source: A radiation source that, if not managed safely or securely, could lead to the death or permanent injury of individuals in a short period of time.

Category 2 source: A radiation source that, if not managed safely or securely, could lead to the death or permanent injury of individuals who may be in close proximity to the radioactive source for a longer period of time than for Category 1 sources.

Category 3 source: A radiation source that, if not managed safely or securely, could lead to the permanent injury of individuals who may be in close proximity to the source for a longer period of time than Category 2 sources. Sources in Category 3 could, but are unlikely to, lead to fatalities.

Chronic radiation exposures: Radiation exposures that occur over extended periods of time (e.g., months to years). Exposure to natural background is a chronic radiation exposure.

Collective dose: The sum of all radiation exposures received by all members of a specified population.

Compensatory measures: Measures beyond those required under existing U.S. NRC regulations. Through security orders, U.S. NRC required some licensees to implement compensatory security measures as conditions of their licenses.

Curie (Ci): A unit of measure for activity equal to 3.7×10^{10} (37 billion) disintegrations (radioactive decays) per second.

D-T reaction: See fusion.

Decay product: A resultant particle from a radioactive disintegration.

Depleted uranium: Uranium from which much of the uranium-235 has been removed.

Deterministic health effect: A tissue reaction that increases in severity with increasing dose, usually above a threshold dose, in affected individuals (organ dysfunction, fibrosis, lens opacification, blood changes, and decrease in sperm count). These are events caused by damage to populations of cells, hence the presence of a threshold dose.

Dose: See *Radiation dose*.

Dose rate: See *Radiation dose rate.*

Effective dose: The equivalent dose averaged over all organs that accounts for the varying sensitivity of different organs and tissues to the biological effects of ionizing radiation. The effective dose has the same units as the equivalent dose.

Equivalent dose: The absorbed dose averaged over the organ or tissue of interest multiplied by a radiation-weighting factor, w_R, to account for the differences in biological detriment (harm) to an organ that result from differences in radiation type and energy for the same physical dose received by the organ. The SI unit of equivalent dose is sievert (Sv); the old unit is the rem. For x rays, gamma rays, and electrons, w_R is 1; for protons, it is 5, for alpha particles, 20; and for neutrons, it ranges from 5 to 20 depending on neutron energy.

Exposure: A metric based on the ability of photons to ionize air. Its old unit roentgen (R) is defined as charge of 2.58×10^{-4} C produced per kilogram of air. The SI unit of exposure is 2.58×10^{-4} C per kilogram of air.

External exposure: An exposure received from a source of ionizing radiation outside of the body (NCRP, 2001).

External cost: A cost from an action or economic transaction that is not included in the monetary cost of the activity or transaction and therefore is borne by parties not directly involved in the transaction.

Fission: The splitting of a nucleus into at least two fragments, accompanied by the release of neutrons and energy. Fission of a nucleus may be initiated by absorption of a neutron or, in some materials such as californium-252, can happen spontaneously.

Fusion: The joining together of two or more nuclei. The most commonly used fusion reaction is the deuterium-tritium reaction, also called the D-T reaction.

Gamma ray: High-energy electromagnetic radiation. In this report, radiation emitted by decay of a radionuclide is always referred to as gamma radiation to distinguish it from radiation from an x-ray generator.

Genetic or hereditary effects are radiation-induced mutations to an individual's genes and DNA that can contribute to the birth of defective descendants.

Graft versus host disease (GVHD): A rare but usually fatal complication of transfusion in which functional donor immune cells (T lymphocytes) attack the recipient's tissues and the recipient's immune system is unable to eliminate the donor lymphocytes.

Greater-than-Class-C waste: Radioactive waste that contains concentrations of certain radionuclides above the Class C limits in 10 CFR § 61.55.

Ground shine: Radiation exposure from material deposited on the ground.

Half-life: The time during which one-half of a given quantity of a radionuclide undergoes radioactive decay.

Half thickness: The thickness of a slab material that reduces by half the intensity of radiation incident on one side of the slab.

Hazard: A potential source of a negative consequence or harm.

High-*Z* material: See atomic number.

Hydrocarbon: In the context of this report, oil or natural gas.

Irradiation: Exposure to radiation.

Increased Controls: A set of security measures required by the U.S. Nuclear Regulatory Commission of Category 1 and 2 radiation source or device licensees.

Ingestion: Uptake of a material into the body via the digestive tract.

Inhalation: Uptake of a material into the body via the respiratory tract.

Internal exposure: An exposure received from a source of ionizing radiation inside of the body (NCRP, 2001).

Ionizing radiation: Radiation that is sufficiently energetic to ionize the matter (i.e., remove electrons from the atoms) through which it moves.

Late effects of radiation: Delayed effects such as fibrosis, atrophy, ulceration, stenosis, or obstruction of the intestine. Late effects may be generic and caused by absorption of radiation directly in the target tissue, or consequential to acute damage in overlying tissues such as mucosa or the epidermis.

Latent cancer: Cancerous lesions in a living organism that have not yet progressed to a stage to be detectable.

Lethal damage: Radiation damage to mammalian cells is divided into three categories: Lethal Damage is irreversible, irreparable, and leads to cell death. Sublethal damage to cells can be repaired in hours unless additional sublethal damage is added that eventually leads to lethal damage. Potentially lethal damage to cells can be manipulated by repair when cells are allowed to remain in a nondividing state.

Natural background radiation: Radiation that exists naturally in the environment. It includes cosmic and solar radiation, radiation from radioactive materials present in rocks and soil, and radioactivity that is inhaled or ingested.

Nondestructive testing (NDT): Testing that does not destroy the object under examination.

Offsite Source Recovery Project: An effort by the National Nuclear Security Administration to recover and secure radiation sources that may pose a danger to public health, safety, and security. The project is run by Los Alamos National Laboratory.

Panoramic irradiator: An irradiation device that does not have shielding built into the device. In such devices, the sources must be housed in thick, shielded structures.

Radiation dose: The quantity of radiation energy deposited in an object or medium divided by the mass of the object or medium. The radiation dose of interest in this report is ionizing radiation. Ionizing radiation doses can be expressed as an absorbed dose, equivalent dose, or effective dose. Its SI unit, gray (Gy), is defined as 1 joule (J) of energy absorbed per kilogram of absorbing medium; its old unit is the rad defined as 100 erg of energy absorbed per gram of absorbing medium.

Radiation dose rate: The quantity of ionizing radiation absorbed by a medium per unit mass of the medium per unit time.

Radiation exposure: The act of being exposed to radiation. Also referred to as *irradiation*. Formally in radiation detection and measurement, radiation exposure is related to the ability of photons to ionize air.

Radiation source: Radioactive material packaged to use the radiation it emits.

Radioactive: Elements that are unstable and transform spontaneously (i.e., decay) through the emission of ionizing radiation, a process known as *radioactive decay*.

Radioactive decay: See *Radioactive*.

Radiography: The use of radiation to create images of a subject, especially the internal features of a subject. Medical radiography is familiar from routine dental examinations. Industrial radiography is a form of nondestructive testing for aircraft wings, pipes, turbines, reinforced concrete construction, and other applications.

Radiological dispersal device (RDD): A device used to spread radioactive material for malevolent purposes. The objective of such a device might be to cause social disruption (panic, evacuation), acute physical harm, the potential for physical harm from chronic exposure, and/or economic damage. An area-denial RDD is one intended to cause contamination that prevents occupation of the contaminated area for an extended period of time.

Radiological exposure device (RED): A device used to cause direct radiation exposure for malevolent purposes.

Radionuclide: An atom with an unstable nucleus, which undergoes radioactive decay.

Radiotherapy: Treatment of disease with ionizing radiation.

Radiosurgery: Focal irradiation techniques that use multiple, non-coplanar radiation beams to deliver a prescribed dose of radiation to a lesions, primarily in the brain.

Resuspension inhalation: Inhalation of radioactive materials that were deposited onto the ground and later resuspended in air.

Risk: As used in this report, the potential for an adverse effect from the accidental or intentional misuse of a radiation source. This potential can be estimated quantitatively if answers to the following three questions can be obtained: (1) What can go wrong? (2) How likely is it that something will go wrong? And (3) What are the consequences? Risk can be expressed in absolute terms or in comparison to other types of risks.

Safety: In the context of this report, concerning prevention of failure, damage, human error, and other inadvertent acts involving radiation sources that could result in accidental radiation exposures.

Safety risks: In the context of this report, risks that arise from exposures of people to radiation as a direct result of accidents involving radiation sources.

Security: In the context of this report, concerning protection against theft, sabotage, and other malevolent acts involving radiation sources.

Self-contained irradiator: "An irradiator in which the sealed source(s) is completely contained in a dry container constructed of solid materials, the sealed source(s) is shielded at all times, and human access to the sealed source(s) and the volume(s) undergoing irradiation is not physically possible in its designed configuration," according to the American National Standards Institute (ANSI) Standard N433.1, "Safe Design and Use of Self-Contained, Dry Source Storage Gamma Irradiators (Category I)." Also called a self-shielded irradiator.

SI: International System of Units (from the French Système International d'Unités), also sometimes referred to as the metric system.

Social cost: Costs to society, including direct and indirect costs paid as money and undesirable effects that are not readily monetized..

Societal risk: All risks that affect society, including the health and safety risks and social risks discussed in this report.

Solubility: The ability of a substance to dissolve in water or, more generally, in a solvent.

Somatic health effect: The harm that exposed individuals suffer during their lifetime, such as radiation-induced cancers (carcinogenesis), sterility, opacification of the eye lens, and life shortening.

Special form radioactive material. Defined in 10 CFR Part 71 as radioactive material that exists as a single solid piece or is encapsulated material that meets certain other requirements.

SPWLA: Society of Petrophysicists and Well Log Analysts.

Stochastic effect: One in which the probability of occurrence increases with increasing dose but the severity in affected individuals does not depend on the dose (radiation carcinogenesis and genetic effects). There is no threshold dose for effects that are truly stochastic, because these effects arise in single cells and it is assumed that there is always some small probability of the effect occurring no matter how small the radiation dose is.

Total Body Radiation syndrome: The response of an organism to acute total body radiation exposure is influenced by the combined response to radiation of all organs constituting the organism. Depending on the actual total body dose above 1 Gy, the response is described as a specific radiation syndrome:

- **Bone marrow syndrome:** 1 Gy < Dose < 10 Gy.
- **Gastrointestinal syndrome:** 10 Gy < Dose < 100 Gy.
- **Central nervous system syndrome:** Dose > 100 Gy.

Transuranic waste: Radioactive waste containing long-lived radioactive transuranic elements (elements with atomic numbers greater than 92) such as plutonium in concentrations greater than 100 nanocuries per gram.

Ultrasonics: The use of high-intensity acoustic energy for materials examination.

Vitrification: A process for immobilizing radioactive material in glass matrixes.

Well logging: The practice of measuring the properties of the geologic strata through which a well has been drilled and recording the results as a function of depth.

X-ray: High-energy electromagnetic radiation. In this report, radiation emitted by a machine such as an x-ray tube or an electron accelerator with a high-*Z* target is always referred to as x-ray radiation to distinguish it from radiation from decay of a radionuclide.

APPENDIX D

INFORMATION-GATHERING MEETINGS

Following is a list of presentations received by the committee during its information-gathering meetings, which were open to the public and included opportunities for public comment.

Meeting 1: July 7, 2006, Washington, D.C.

NRC Perspectives on the Radiation Source Use and Replacement Study.

Brian Sheron, Director of Nuclear Regulatory Research, U.S. Nuclear Regulatory Commission (U.S. NRC)

Congressional Perspectives on the Radiation Source Use and Replacement Study.

Michal Freedhoff, Senior Policy Associate, Office of Representative Edward J. Markey

U.S. NRC Studies on Reducing Source Dispersability.

Makuteswara Srinivasan, Materials Engineering Directorate, Office of Nuclear Regulatory Research, U.S. NRC

U.S. NRC Interim Inventory and Source Tracking System.

Patricia K. Holahan, Deputy Director Division of Industrial and Medical Nuclear Safety, U.S. NRC

EPA Work on Alternatives to Radioactive Sources.

Sally Hamlin, Radiation Protection Division, U.S. Environmental Protection Agency (U.S. EPA)

FDA's Role in Regulation of Radiation Generating Devices and Irradiated Medical Products.

Orhan Suleiman, Senior Science Policy Advisor, Center for Drug Evaluation and Research, Food and Drug Administration (U.S. FDA)

Radiation Source Manufacturers.

Grant Malkoske, P. Eng. Chairman, the International Source Suppliers and Producers Association (ISSPA) and Gamma Industry Processing Alliance (GIPA)

Meeting 2: September 11–12, 2006, Washington, D.C.

International Atomic Energy Agency Perspectives on Radiation Source Uses and Replacements.

Brian Dodd, Consultant and President of the Health Physics Society

Interagency Report to Congress on Radiation Source Security, Nuclear Regulatory Commission.

Merri Horn, Senior Project Manager, Officer of Nuclear Material Safety and Safeguards, U.S. NRC

Recommendations and Alternative Technologies to IAEA Category 1 and 2 Radiation Sources

Alternative: Technology Subgroup Radiation Source Protection and Security Task Force.

Tony Huffert (U.S. NRC), for the Interagency Task Force Subgroup on Alternative Technologies

Panel Discussion of the Alternatives Subgroup for the Report to Congress.

Tony Huffert, U.S. NRC
Constance Rosser, U.S. FDA
Sally Hamlin, U.S. EPA
Ruth Watkins and Joel Rabovsky, U.S. DOE
Kirsten Cutler and Brendan Plapp, U.S. State Department

Perspectives from the States.

Pearce O'Kelley, Conference of Radiation Control Program Directors
Barbara Hamrick, Organization of Agreement States

Meeting 3: October 26–28, 2006, Houston, Texas

Oil-Well Logging.

Allen Gilchrist, Baker Hues

Practical Considerations in Current Applications for Nondestructive Testing.

Michael Creech, Vice President, Acuren Inspection Inc.

Radiography Methods, Equipment, and Current Practices.

R. D. "Donny" Dicharry, President, Source Production & Equipment Co., Inc.

Radiosurgery.

Chuck Vecoli, Senior Business Marketing Manager, Elekta, Inc.

Comparison of Radiation Processing Technologies.

Mark A. Smith, CHP Vice President, Radiation Services, Sterigenics International

Capabilities of Techniques and R&D for Nondestructive Testing.

Glenn Light, Southwest Research Institute, and Grady Legleder, IHI Southwest Technologies

Industrial Accelerators and X-ray Tubes: Capabilities and Limitations as a Replacement to Isotropic Sources.

Lester Boeh, Varian Medical Systems

Houston Medical Center tour.

Karl Prado, Patricia Eifel, Benjamin Lichtiger, Shiao Woo, Michael Gillin, Ann Lawyer, Radhe Mohan, Ray Meyn, Jay Poston, Peggy Tinkey, and Almon Shiu of MD Anderson Cancer Center

Otto Zeck of Memorial Hermann Hospital

National Center for Electron Beam Food Research Facility Tour.

Mickey Speakmon and Les Braby, National Center for Electron Beam Food Research, Texas A&M University, College Station

Meeting 4: December 8–9, 2006, Washington, D.C.

J. L. Shepherd and Associates on Cesium Chloride and Irradiators.
Wayne Norwood, for J.L. Shepherd and Associates

Rad Source Technologies, Inc. on X-Ray Blood Irradiator.
Randol Kirk, President, Rad Source Technologies, Inc.

ViewRay IGRT with MRI and Co-60 Sources.
Jim Dempsey, CSO, ViewRay, Inc.

Nordion-NOMIS CT-Co-60 IMRT Venture, X-Ray Blood Irradiator.
Mark Vist, MDS-Nordion

IBA Industrial X-Ray and E-Beam Irradiation Devices.
Marshall Cleland, Technical Advisor, RDI, Member of the IBA Group

Meeting 5: February 1–2, 2007, Irvine, California

U.S. Offsite Source Recovery Program.
Mike Pearson, Los Alamos National Laboratory

Discussion on Cesium Chloride, Cesium Glass, and Cobalt Irradiators.
J. L. Shepherd, J.L. Shepherd and Associates